T0191862

Analog Circuits and Signal Processing

More information about this series at http://www.springer.com/series/7381

Marijn van Dongen • Wouter Serdijn

Design of Efficient and Safe Neural Stimulators

A Multidisciplinary Approach

Marijn van Dongen
NXP Semiconductors N.V.
Nijmegen, The Netherlands

Wouter Serdijn
Delft University of Technology
Delft, The Netherlands

/

ISSN 1872-082X ISSN 2197-1854 (electronic)
Analog Circuits and Signal Processing
ISBN 978-3-319-80278-7 ISBN 978-3-319-28131-5 (eBook)
DOI 10.1007/978-3-319-28131-5

Springer Cham Heidelberg New York Dordrecht London

Softcover reprint of the hardcover 1st edition 2016

Printed on acid-free paper

Springer International Publishing AG Switzerland is part of Springer Science+Business Media (www.springer.com)

En dit is de reeden, dewelke my beweegt, om eenige experimenten, die ik al over lang omtrent deeze zaak gedaan heb, in het ligt te geeven, en alzoo ik die van een zeer groote consequentie en gewigt oordeel, zoo zou ookmyn verzoek zyn, omdie ernstig te willen naadenken, en op den toetsteen der waarheid te stellen

Jan Swammerdam,
Bybel der natuure, 1737

Preface

Electrical neural stimulation is an established treatment methodology for an increasing number of neural pathologies, while its application is under consideration for an even larger number of diseases. In line with this development, there is a need for neural stimulator devices that are safe and reliable and have a small form factor. The design of such devices requires a multidisciplinary approach, combining the needs from neurological, physiological, electrochemical, and electrical perspectives.

The concept of electrical stimulation is typically approached from two distinct directions. One way starts at the neuron and asks the question: what kind of signals are needed to achieve the desired neural modulation? This approach typically calculates the neural response based on the electrode configuration, the applied electrical field, and the physical properties of the neurons under consideration. The other approach starts at the stimulator and asks the question: what kind of circuit techniques can be used to implement the stimulation signals? Typical characteristics addressed are power efficiency, safety aspects (such as charge cancelation), and scalability (such as number of outputs).

Both approaches seem to operate rather isolated from each other. The first approach is typically unaware of how a certain optimal waveform can be translated into electrical circuitry. Similarly, the second approach is often not aware how alternative circuit topologies will influence the neural activation mechanism.

It would be much better to combine both approaches: what kind of signals allow for both efficient activation and efficient circuits? This book aims to assist neural stimulator circuit designers in taking such an approach by presenting the complete stimulation sequence: from the neural stimulator down to the neuronal membrane where the activation (or inhibition) takes place. By understanding this complete chain, it is possible to devise novel stimulator architectures while understanding the safety aspects that are important for neural stimulation. Some examples of novel approaches are given throughout the book. These include considerations about safety, electrochemical stability, and stimulator architectures.

One of the drawbacks of taking a fundamentally different approach is that it is generally much harder to get the work recognized in the scientific community. The following story illustrates this point well and is therefore worth sharing. In 1892 the scientist Jan Leendert Hoorweg from Utrecht, the Netherlands, published a bold article in the journal *Archiv für die gesamte Physiologie des Menschen und der Tiere* (the contemporary *Pflügers Archiv*) [1]. He had studied the conditions under which a charged capacitor could excite muscle contractions in human subjects. He found that the relationship as proposed by the famous founding father of electrophysiology, Emil du Bois-Reymond, seemed invalid. In 1845 E. du Bois-Reymond had formulated a relationship [2] in which the momentary muscle movement $\epsilon(t)$ was hypothesized to depend on the momentary change in stimulation current: $\epsilon(t) = F[di(t)/dt]$.

Hoorweg, not being satisfied with the empirical "evidence" from [2], conducted a series of systematic experiments and found a relationship independent on $di(t)/dt$, but on the stimulation circuit parameters used such as capacitance, resistance, and voltage. His fundamentally different point of view caused big consternation, and many famous scientists, such as Eduard Pflüger, were quick to reject his idea in a single-sentence publication without any further proof [3].

It took another 9 years until, in 1901, Georges Weiss established a relationship between stimulation charge and duration [4] and showed that the measurements from Hoorweg were actually correct. In 1909 Louis Lapicque reformulated [10] (Chap. 2) the results into the famous strength-duration curve, which is now one of the fundamental principles of neural stimulation.

Discovering Hoorweg's story gave me an odd sense of satisfaction, not only because it turned out that his ideas were correct but mainly because it showed me that convincing the scientific community to consider alternative approaches was as difficult as it is today. During my years of research, I also experienced that it is not always easy to convince the society to at least allow alternative ideas as an input to the field.

It is thanks to the people around me that I was able to push on to continue and prove the validity and usefulness of the ideas and concepts that are presented in this book. In this aspect I would like to thank the section Bioelectronics of the Delft University of Technology: it was a privilege to be part of this group of people. Furthermore, I enjoyed great cooperation with several other research groups as part of the SINs consortium in which I experienced the multidisciplinary character of this research field. Here I would like to mention the Neuroscience Department of Erasmus University Rotterdam and the Neurosurgery Department of the University of Otago and the University of Antwerp.

Finally, I would like to thank the most important people in my life: my wife Lin and daughter Danya. It is you who gave me the strength and support needed to finish my work and this book.

Nijmegen, The Netherlands Marijn van Dongen
October 2015

References

1. Hoorweg, J.L.: Ueber die elektrische Nervenerregung. Arch. Gesame. Physiol. Menschen Tiere **52**(3–4), 87–108 (1892)
2. du Bois-Reymond, E.: Untersuchungen über thierische elektricität. In: Von dem allgemeinen Gesetze der Nervenerregung durch den elektrischen Strom (Band 1, Chapter 2.2). G. Reimer, Berlin (1848)
3. Pflüger, E.: J.L. Hoorweg und die electrische Nervenerregung. Arch. Gesame. Physiol. Menschen Tiere **53**(11–12), 616 (1893)
4. Weiss, G.: Sur la possibilité de rendre comparables entre eux les appareils servant à l'excitation électricque. Arch. Ital. Biol. **35**(1), 413–446 (1901)

Contents

Chapter 1
Introduction

Abstract This chapter introduces the reader to the concept of neural stimulation and electrical neural stimulation in particular. It shortly discusses various types of stimulation and the possible clinical applications. It then illustrates the technological challenges for the design of neural stimulator based on a case study of a spinal cord stimulator. Several of these challenges will be addressed in the subsequent chapters of which an overview is given at the end of this chapter.

1.1 Neural Stimulation

In 1658 the Dutch Scientist Jan Swammerdam performed the first documented neuromuscular physiology experiment: he described that the muscle of a dissected frog contracts when the associated nerve was irritated [1]:

> ... zoo vat men de Spier aan weerzyden by zyne peezen, en als man dan de neerhangende Senuw met een schaarken of iets anders irriteert, zoo doet men de Spieren zyn voorige en verloore beweeging weer harhaalen[1]

More than a century later half of Europe was being entertained by the "resurrection" of dead animals and even humans using electrical shocks [2]. This was due to the legendary experiment by Luigi Galvani that led to the discovery of bio-electricity and inspired Alessandro Volta to invent chemical electricity in the form of a battery [3]. These discoveries were instrumental towards the development of electrophysiology [4] and our understanding of the nervous system.

As will be seen in Chap. 2, the nervous system is an electrochemical system. The traditional means to treat diseases use medicine (drugs) and many of them tap in on the chemical component of the nervous system, e.g., by influencing the gating of specific ion channels in neurons [5]. Drawback of this approach is that drugs often influence the whole body and are therefore likely to introduce various unwanted side effects.

[1]One suspends the muscle at both tendons and once the nerve is irritated with scissors or some other tool, the muscle regains its lost movements.

M. van Dongen, W. Serdijn, *Design of Efficient and Safe Neural Stimulators*,
Analog Circuits and Signal Processing, DOI 10.1007/978-3-319-28131-5_1

Neural stimulation on the other hand taps in on the electrical component of the nervous system: it is possible to artificially generate or block action potentials in a predefined area. This means that neural stimulation has the potential to act in a more localized manner. Furthermore, the electrical component generally has a virtually instantaneous response: the effects of neural stimulation are usually immediately noticeable upon activation and, maybe even more importantly, are reversible when stimulation is disabled. This in contrast to the chemical component, which usually takes much more time to settle.

The programmability of the neural stimulation device allows for automatic adjustment of the stimulation parameters (i.e., the dosage): a feedback loop can be established that tailors the stimulation to the subject's needs. Although promising, closed-loop operation currently has a limited amount of clinical applications [6]. Most often manual tuning is used, in which the parameters are adjusted by a physician or the subject based on the empirical response. This is very similar to dose adjustments in drug therapy. This book is mainly concerned with the design of the neural stimulator itself and hence the design of the feedback loop is not treated extensively.

Neural stimulation can use a variety of modalities. This book focuses on electrical stimulation, in which an electrical current through electrodes is used to generate a potential difference in the target tissue that will establish the desired neural recruitment. Other common stimulation modalities include magnetic stimulation (e.g., transcranial magnetic stimulation (TMS), which is non-invasive, but compromises in selectivity and portability [7, 8]), light (e.g., optogenetic stimulation, which has excellent selectivity, but requires genetic modification of the target tissue [9]), and sound (ultrasound stimulation [10, 11]).

Electrical stimulation has shown very impressive results in clinical practice. Spinal cord stimulation (SCS) has been successfully applied for pain suppression [12], motor disorders [13], and bladder control [14]. Vagus nerve stimulation (VNS) is targeting a specific nerve and is a therapy that is used to treat epilepsy [15] and depression [16]. Furthermore various other potential applications have been identified, such as the treatment of tinnitus (phantom sound perception) [17].

Deep brain stimulation (DBS) has become a widely used stimulation technique that positions electrodes in various target regions in the brain itself [18, 19]. It has been used to treat diseases such as essential tremor, Parkinson's disease, dystonia, Tourette syndrome, pain, depression, and obsessive compulsive disorder. Electrical stimulation can furthermore help to restore sensory input: a cochlear implant stimulates the auditory nerve to restore hearing [20], a retinal implant stimulates the retina to restore vision [21], and vestibular implants have the potential to treat vestibulopathy by restoring the sense of balance [22].

The success of these techniques can be largely attributed to the technological progression that was made during the last century. Many of the stimulator devices use Integrated Circuit (IC) technology, are made of bio-compatible materials, and use advanced battery and/or power harvesting technology. To illustrate these components, a closer look is given on a stimulator device as used in clinical practice.

1.2 Case Study: SCS Device

Let's take as an example a spinal cord stimulator device, typically referred to as an implantable pulse generator (IPG). In Fig. 1.1 the in- and outside of such a device, which is typically implanted in the chest or the abdomen, are shown. The casing is made from a bio-compatible material such as titanium. Furthermore it features a connector for the cables that lead towards the electrodes in the target area. These types of devices are typically controlled (and often also charged) from the outside via an inductive link that is established using the power antenna.

The inside of the IPG shows the printed circuit boards (PCBs) with the electronics and the battery. The battery takes up a remarkable amount of space in the package. This is on one hand due to the relatively high power that is used for SCS applications (stimulation amplitudes of up to 25.5 mA as stated in the technical specifications [23]). On the other hand, the power is determined by the efficiency of the operation of the IPG, such as the stimulation frequency.

Another remarkable component on the PCB that takes up a significant amount of space are the coupling capacitors. These capacitors are included for safety reasons and provide AC coupling of the electrodes to the stimulator. Various studies have focused on eliminating these capacitors [24], although it is unclear whether the alternatives offer the same level of safety.

One of the aspects that would improve the performance of this particular IPG mostly is the size of the device. A smaller device would reduce the impact on implantation and allows the IPG to be placed closer to the electrodes, reducing or

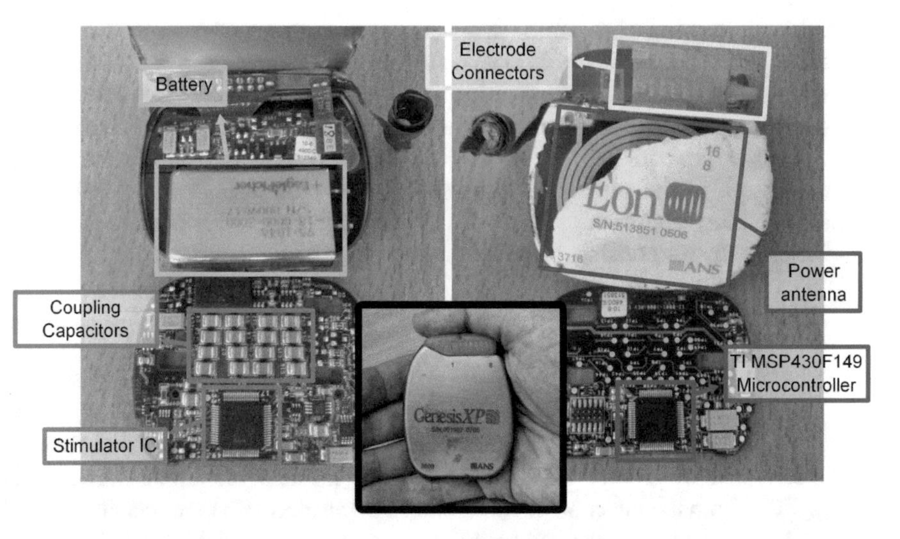

Fig. 1.1 Pictures of IPGs used for spinal cord stimulation. In the center an explanted ANS (currently St. Jude Medical) Genesis IPG is shown that uses a primary cell for powering the device. The other pictures show an opened ANS (currently St. Jude Medical) EON IPG featuring a rechargeable battery. Remarkable is the size required for the battery and the coupling capacitors

eliminating the electrode cables, which are a common source of device failure [25]. Reducing the size can be obtained by reducing the power consumption (allowing for a smaller battery) or achieving a higher level of integration (smaller number of discrete components).

Closed-loop stimulation is an effective way to reduce the stimulation frequency and hence power consumption, because the stimulation pattern tailors itself to the subject, reducing the amount of "overstimulation." This is illustrated by clinical closed-loop epilepsy suppression, which reports an average active time of just 5 min per 24 h [6]. Besides these improvements on the functional level, improvements in the electrical design can achieve a higher efficiency, but also a higher level of integration. This book focuses on improving the electrical design of neural stimulators.

1.3 Goal of This Book

All applications of electrical stimulation need their own specialized stimulator design, electrode configuration, and stimulation parameters [26]. However, the underlying stimulation circuitry of these devices is often surprisingly similar: a constant current or voltage source is imposed on the electrodes for a certain amount of time to achieve the desired neuronal response. Most of the design efforts improving the efficiency and integration of the system focus on the electrical engineering aspects exclusively and assume a constant voltage or current source is the required stimulation paradigm.

Although these design efforts have led to significant improvements in the electrical performance of the stimulator devices, it has not been shown that the current paradigm necessarily leads to the most efficient and safe stimulation. This book takes one step back and combines the electrophysiological and electrochemical principles that govern FES with the electrical engineering principles that are used for the design of the stimulation devices. The question is how we can use this multidisciplinary approach to improve the performance of FES devices in terms of efficiency (by introducing neural recruitment strategies) and safety.

1.3.1 Neural Recruitment Strategies

The first stimulation strategy introduced is high frequency duty cycled stimulation. The electrophysiological principles that govern the artificial recruitment of neurons during FES, from the injection of current through the electrode up to the generation of an action potential on the cell membrane, are used to validate that this type of stimulation is effective. The principle is realized with circuit topologies, commonly used in power electronics, to implement a system that achieves, among other advantages, power efficient operation. The system architecture completely abandons the idea of a (constant) current or voltage source and uses a different output topology that aims to achieve the same kind of activation in the target area as the classical circuits.

The second stimulation strategy that is introduced moves the gain in efficiency from the electrical circuit towards the electrophysiology: the circuit uses a voltage or current based output, but implements arbitrary waveform capabilities to allow the user to select the most efficient stimulation waveform. The system is designed to be used in a multimodal stimulation experiment with animals to treat tinnitus. One of the challenges in such an arbitrary waveform stimulator design is to guarantee the (electrochemical) safety.

1.3.2 Safety Aspects

The safety aspect is considered by introducing circuit techniques that investigate the electrochemical processes that occur at the electrode–tissue interface. The established approach to prevent harmful electrochemical reactions is to implement a charge balanced biphasic stimulation waveform and by using coupling capacitors. By understanding the electrochemical processes, circuit techniques are used to evaluate the current safety mechanisms and new options are considered.

Throughout the book, the proposed designs and mechanisms do generally not focus on one particular application, but instead aim to be applied in a more general fashion. However, most findings are verified using measurements that use a certain type of electrodes or measurement setup that is typical for a specific application.

1.4 Outline of This Book

This book is composed out of two parts. In the first part the focus will be on the electrochemical and electrophysiological principles that are used in the proposed designs. The goal is to introduce novel aspects that can improve neural stimulation in terms of safety and efficiency. In the second part the principles introduced in the first part will be used in the electrical design of neural stimulators.

The first part starts in Chap. 2 with a review of the basic electrophysiological and electrochemical principles that will be used in the subsequent chapters. The focus will be on modeling of the physical processes that describe the neural stimulation process which influences the neural activity due to the injection of current through electrodes. This chapter is highly multidisciplinary and uses principles from neurophysiology, electrochemistry, electrical field calculations, and electrical circuit theory. All these aspects play a role in neural stimulation and each of these aspects will be used in the subsequent chapters.

Safety is of major concern for neural stimulator devices, since a faulty stimulation signal can cause irreversible damage to both the tissue as well as the electrodes.

Chapter 3 uses the electrical models for the electrode–tissue interface to treat several safety aspects related with the prevention of harmful electrochemical processes. It discusses the consequences of using coupling capacitors between the stimulator and the electrodes. Furthermore it introduces a stimulation design using a feedforward control mechanism that aims to bring the interface back to equilibrium after a stimulation cycle.

Chapter 4 switches gears to the efficiency of neural stimulation. It explores the use of a fundamentally different stimulation paradigm: instead of using a voltage or current source to drive the electrodes, a high frequency switched-mode stimulation signal is used. Switched-mode operation is a common technique to improve the power efficiency of amplifiers (e.g., class D operation) and a neural stimulator could benefit in a similar way. This chapter investigates whether a switched-mode stimulator output stage is able to recruit neurons in a similar way as when a classical stimulation pattern is used. Using in vitro measurements on brain slices the efficacy of switched-mode stimulation is verified.

The second part starts with Chap. 5 which serves as an introduction to the electrical design of neural stimulators. Several system design aspects of neural stimulators are discussed and their relative importance in various applications is outlined. Furthermore, the chapter links each aspect to the two subsequent chapters, which both describe the design of a neural stimulator system. Both systems have been designed with different design goals and hence their system requirements are completely different.

This first stimulator system (discussed in Chap. 6) features the design of an arbitrary waveform stimulator. The system uses a classical voltage or current based output, but the waveform can be fully customized by the user, while the safety is guaranteed by several feedback mechanisms. This kind of system is particularly interesting for neuroscientific experiments in which novel stimulation designs are explored in order to evoke the desired response. This system is realized in discrete form to be applied in animal experiments for the treatment of tinnitus in which special burst stimulation waveforms are part of a multimodal stimulation paradigm that aims to evoke reconditioning of the nervous system.

The second stimulator system is treated in Chap. 7 and uses the high frequency switched-mode stimulation strategy that was explored in Chap. 4. The resulting system features 8 independent stimulation channels that connect to any of the 16 electrodes at the output. The system improves existing neural stimulator systems in terms of power efficiency (especially if current steering techniques are used) and the number of external components (which improves both safety as well as system size). The system also features comprehensive digital control, which allows the system for operation in a stand-alone manner.

The structure of this book underlines the multidisciplinary approach that is required in this field: only by combining the understanding of the neurophysiological principles that form the basis of neural stimulation with the electrical engineering design skills to design neural stimulation circuits, it is possible to propose novel stimulation strategies that can improve on aspects such as safety and efficiency.

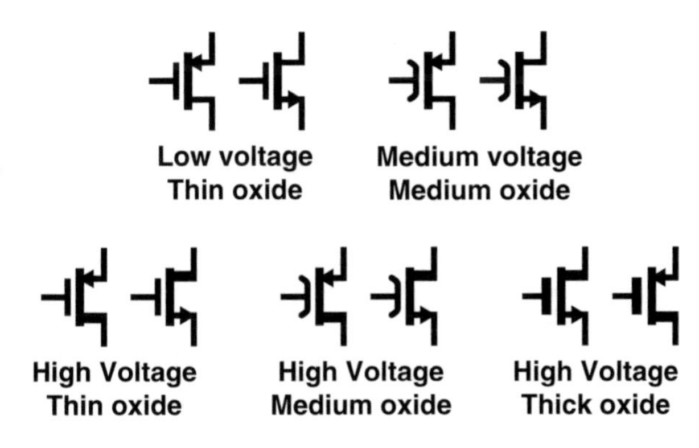

Fig. 1.2 Symbols used in this book for different types of low, medium, and high voltage transistors that are available in the IC design kits. The levels of isolations can apply to the drain terminal and/or the gate terminal, although the technology in Chap. 6 only offers thin gates. The low voltage/thin oxide transistors are the standard transistors for the technology

1.5 Symbols

In the second part of this book the IC design uses technologies that offer high voltage DMOS transistors. These transistors are needed for neural stimulators, because the output voltage often exceeds the breakdown voltage of the standard transistors. In Fig. 1.2 the symbols that are used for these devices are depicted. The breakdown voltage is categorized by three levels: low voltage (standard device rating), medium voltage, and high voltage.

The technology used in Chap. 6 offers isolation that applies to the drain only: the gate voltage is always limited to the normal operating voltage. The technology in Chap. 7 also allows for higher gate voltages by offering thicker gate oxides. All possible and available combinations are shown in Fig. 1.2.

References

1. Swammerdam, J.: Proefnemingen van de particuliere beweeging der spieren in de kikvorsch, die in het gemeen op alle de bewegingen der spieren in de menschen en beesten toegepast worden. In: Swammerdam, J., Boerhaave, H. (eds.) Bybel der natuure, deel II, Leiden (1737)
2. Ashcroft, F.: Spark of Life – Electricity in the Human Body. W.W. Norton and Company, New York (2012)
3. Geddes, L.A., Hoff, H.E.: The discovery of bioelectricity and current electricity. The Galvani-Volta controversy. IEEE Spectr. **8**(12), 38–46 (1971)
4. Verkhratsky, A., Krishtal, O.A., Petersen, O.H.: From Galvani to patch clamp: the development of electrophysiology. Pflügers Arch. **453**(3), 233–247 (2006)
5. Kandel, E., Schwartz, J.H., Jessell, T.: Principles of Neural Science. McGraw-Hill, New York (2000)
6. Sun, F.T., Morrell, M.J.: Closed-loop neurostimulation – the clinical experience. Neurotherapeutics **11**(3), 553–563 (2014)
7. Hallett, M.: Transcranial magnetic stimulation and the human brain. Nature **406**, 147–150 (2000)
8. Rossini, P.M., Rossi, S.: Transcranial magnetic stimulation: diagnostic, therapeutic and research potential. Neurology **68**(7), 484–488 (2007)

9. Fenno, L., Yizhar, O., Deisseroth, K.: The development and application of optogenetics. Annu. Rev. Neurosci. **34**, 389–412 (2011)
10. Tyler, W.J., Tufail, Y., Finsterwald, M., Taumann, M.L., Olson, E.J., Majestic, C.: Remote excitation of neuronal circuits using low-intensity, low-frequency ultrasound. PLoS One **3**(10) (2008). http://journals.plos.org/plosone/article?id=10.1371/journal.pone.0003511
11. Menz, M.D., Oralkan, Ö., Khuri-Yakub, P.T., Baccus, S.A.: Precise neural stimulation in the retina using focused ultrasound. J. Neurosci. **33**(10), 4550–4560 (2013)
12. Kunnumpurath, S., Srinivasagopalan, R., Vadivelu, N.: Spinal cord stimulation: principles of past, present and future practice: a review. J. Clin. Monit. Comput. **25**(5), 333–339 (2009)
13. Harkema, S., Gerasimenko, Y., Hodes, J., Burdick, J., Angeli, C., Chen, Y., Ferreira, C., Willhite, A., Rejc, E., Grossman, R.G., Edgerton, V.R.: Effect of epidural stimulation of the lumbosacral spinal cord on voluntary movement, standing and assisted stepping after motor complete paraplegia: a case study. Lancet **377**(9781), 1938–1947 (2011)
14. Brindley, G.S., Polkey, C.E., Rushtom, D.N.: Sacral anterior root stimulators for bladder control in paraplegia. Paraplegia **20**, 365–381 (1982)
15. Ben-Menachem, E.: Vagus-nerve stimulation for the treatment of epilepsy. Lancet Neurol. **8**(1), 477–482 (2002)
16. Sackheim, H.A., Rush, A.J., George, M.S., Marangell, L.B., Husain, M.M., Nahas, Z., Johnson, C.R., Seidman, S., Giller, C., Haines, S., Simpson, R.K., Goodman, R.R.: Vagus nerve stimulation (VNS) for treatment-resistant depression: efficacy, side effects, and predictors of outcome. Neuropsychopharmacology **25**, 713–728 (2001)
17. Engineer, N.D., Riley, J.R., Seale, J.D., Vrana, W.A., Shetake, J.A., Sudanagunta, S.P., Borland, M.S., Kilgard, M.P.: Reversing pathological neural activity using targeted plasticity. Nature **470**, 101–104 (2011)
18. Lyons, M.K.: Deep brain stimulation: current and future clinical applications. Mayo Clin. Proc. **86**(7), 662–672 (2011)
19. Perlmutter, J.S., Mink, J.W.: Deep brain stimulation. Annu. Rev. Neurosci. **29**, 229–257 (2006)
20. Niparko, J.K.: Cochlear Implants, Principles and Practices. Lippincott Williams and Wilkins, Philadelphia (2009)
21. Dagnelie, G.: Retinal implants – emergence of a multidisciplinary field. Curr. Opin. Neurol. **25**(1), 67–75 (2012)
22. Merfeld, D., Lewis, R.F.: Replacing semicircular canal function with a vestibular implant. Curr. Opin. Otolaryngol. Head Neck Surg. **20**(5), 386–392 (2012)
23. Eon Rechargeable IPG tech specs. St. Jude Medical Inc (2013). http://professional.sjm.com/products/neuro/scs/generators/eon-rechargeable-ipg#tech-specs. Cited 14 July 2014
24. Liu, X., Demosthenous, A., Donaldson, N.: An integrated implantable stimulator that is fail-safe without off-chip blocking-capacitors. IEEE Trans. Biomed. Circuits Syst. **3**(2), 231–244 (2008)
25. Lanmüller, H., Wernisch, J., Alesch, F.: Troubleshooting for DBS patients by a non-invasive method with subsequent examination of the implantable device. In: Proceedings of the 11th Mediterranean Conference on Medical and Biomedical Engineering and Computing, vol. 11, pp. 651–653 (2007)
26. Albert, G.C., Cook, C.M., Prato, F.S., Thomas, A.W.: Deep brain stimulation, vagal nerve stimulation and transcranial stimulation – an overview of stimulation parameters and neurotransmitter release. Neurosci. Biobehav. Rev. **33**(7), 1042–1060 (2009)

Part I
Towards Safe and Efficient Neural Stimulation

This part introduces the reader to the electrophysiological and electrochemical principles that are important for neural stimulation. The complete stimulation chain from neural stimulator down to the single neuron is discussed. These principles are used in the subsequent chapters to introduce various concepts that have important consequences for both the safety and efficiency of neural stimulation.

Chapter 2
Modeling the Activation of Neural Cells

Abstract This chapter treats a short review regarding the working principles and modeling of neurons. These concepts form the theoretical basis of neural stimulation and are therefore used throughout the book. In the first section the physiological principles of neurons are discussed and an electrical model of the neural membrane will be derived. In the second section the stimulation of neural tissue will be discussed. It will be seen how the passage of current through electrodes will ultimately lead to physiological change in the neural tissue.

2.1 Physiological Principles of Neural Cells

2.1.1 Neurons

Almost all neural cells, or neurons, share four common functions: reception, triggering, signaling, and secretion [1]. In Fig. 2.1 an illustration of these components is given. At the input of the neuron the incoming signals are *received* at the *dendrites* of the neuron. The number and nature of these signals depend on the type of neuron: neurons can have anywhere from a few up to hundreds of thousands of inputs. Input signals are always graded (analog) signals. For the neuron in Fig. 2.1 the input is formed by synaptic potentials, resulting from the neurotransmitter release of other cells.

All the local signals will come together at a point where a *trigger* can be generated if a certain threshold is reached. The trigger is an all-or-none spike shaped signal called an action potential. In the neuron in Fig. 2.1 the trigger point is at the *soma* of the neuron (more specifically the axon hillock). The soma also contains the nucleus of the neuron and synthesizes most neuronal proteins that are essential for operation of the cell.

The action potentials generated at the trigger point will propagate over the *axon* of the neuron. Action potentials propagate over the axon in a regenerative manner without loss of amplitude and therefore the signal can cover large distances. Because the shape and amplitude of the action potential remain constant, the information is contained in the number of spikes and the time between them. The axon therefore has a *signaling* function. An axon may be *myelinated*, which means it is surrounded by myelin sheath, an insulating layer. The myelin is discontinuous and the points

M. van Dongen, W. Serdijn, *Design of Efficient and Safe Neural Stimulators*, Analog Circuits and Signal Processing, DOI 10.1007/978-3-319-28131-5_2

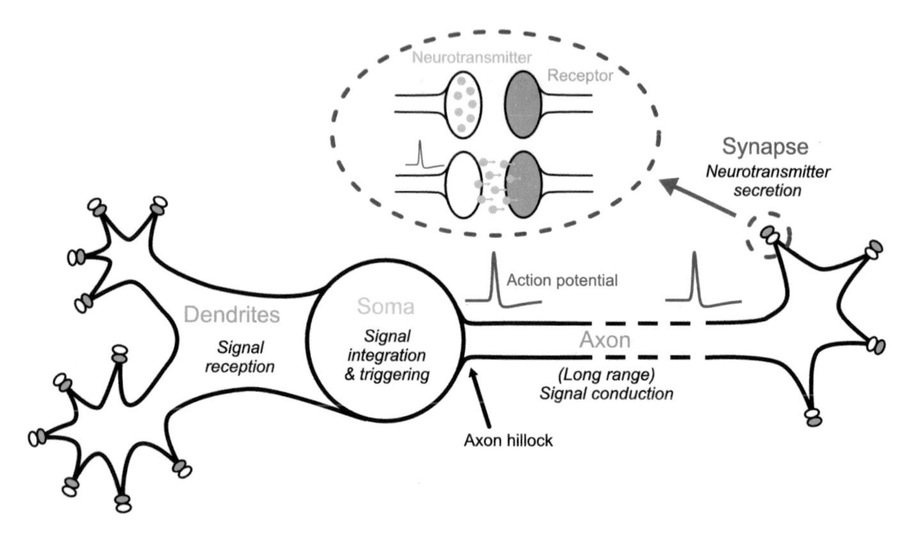

Fig. 2.1 Schematic representation of the basic organization common to all nerve cells: signal reception (dendrites), signal integration and triggering (soma), signal conduction (axon), and neurotransmitter secretion (synapse)

where it is absent are called the *nodes of Ranvier*. Myelin greatly increases the propagation velocity of the axon potentials.

The end of the axon connects to other cells by means of a *synapse*. The action potential causes the *secretion* of a neurotransmitter, which is subsequently received by the receptor, thereby forming a unidirectional connection. The release of a neurotransmitter can be considered the output of the neuron. The neurotransmitter signal is again graded: its amplitude depends on the number and frequency of action potentials that are generated. Whether a synapse has a excitatory or inhibitory effect (increasing or decreasing the chance for an action potential, respectively) depends on the receptor only: one single neurotransmitter can cause both excitatory and inhibitory responses, depending on the receptor.

2.1.2 Modeling of the Cell Membrane

The graded signals at the dendrites as well as the action potentials manifest themselves as changes in the voltage over the cell membrane of the neuron. The membrane consists of two layers of phospholipids that form a seal between the inside of the cell (cytoplasm) and the outside (extracellular fluid). The membrane voltage $V_m = V_{in} - V_{out}$ is defined as the difference between the potentials of the cytoplasm and extracellular fluid. The cytoplasm is characterized by a high concentration of potassium (K^+) and large organic anions (denoted as A^-). The extracellular space has a surplus of sodium (Na^+) and chloride (Cl^-).

The membrane itself is a very good isolator, but diffusion of ions through the membrane is possible through ion channels (proteins that span the membrane). Due to the movement of charge, an electric field forms across the membrane, which will lead to a conduction current opposing the diffusion current. The voltage at which the conduction and diffusion current are in equilibrium for one type of ion is called the Nernst voltage and is given by Malmivuo and Plonsey [2]:

$$V_x = \frac{RT}{z_x F} \ln \frac{c_{i,x}}{c_{o,x}} \tag{2.1}$$

Here R is the gas constant [8.314 J/(mol K)], V_x is the Nernst potential for one specific ion type x, T is the temperature [K], F is Faraday's constant [$9.6 \cdot 10^4$ C/mol], z_x is the valence of the ion, and $c_{i,x}$ and $c_{o,x}$ are the intracellular and extracellular ionic concentrations [mol/cm^3], respectively. Each type of ion channel can now be modeled with a voltage source that is equal to the Nernst potential in series with a conductance $g_{m,x}$ that represents the conductivity of the ion channel. This is depicted in Fig. 2.3 for the sodium, potassium, and chloride channels. Due to the fact that the Nernst potentials of the individual species are unequal, the total membrane is in a dynamic equilibrium: there is a constant flux of ions flowing through the membrane.

The concentration gradients of the sodium and potassium ions are maintained by means of an Na^+-K^+-pump, as also depicted in Fig. 2.2. This is another membrane spanning protein that pumps out 3 Na^+ ions for every 2 K^+ ions that it pumps in the cell. This electrogenic behavior causes V_m to be slightly more negative and can be modeled by two current sources as depicted in Fig. 2.3. The influence of the pump is usually very small and is therefore often neglected in the model.

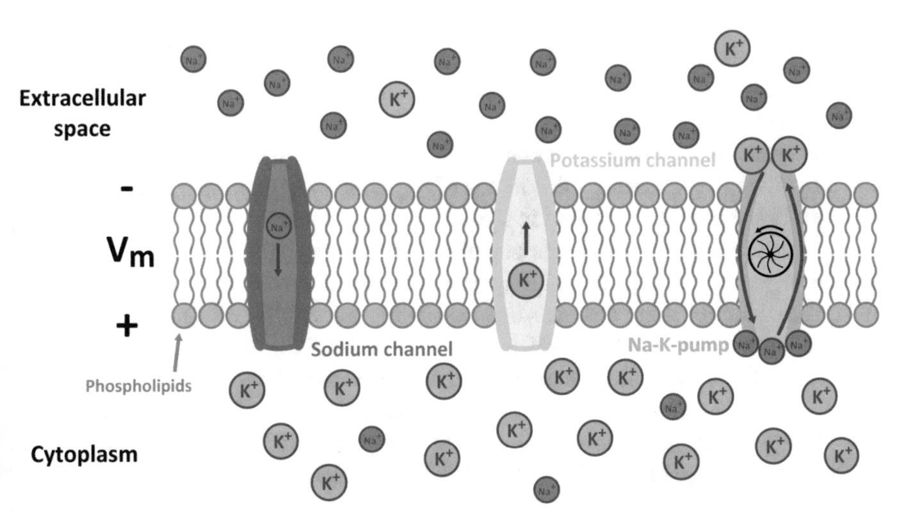

Fig. 2.2 Schematic representation of the cell membrane with ion channels for the most important cations (Na^+ and K^+). The membrane voltage $V_m = V_{in} - V_{out}$ is also depicted. The anions (Cl^- and A^-) are omitted for clarity

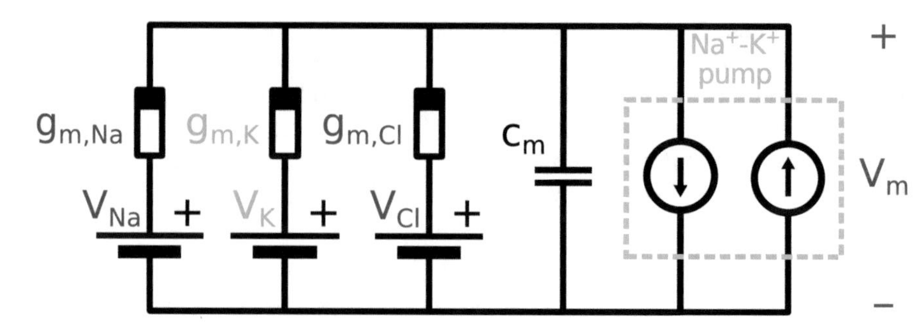

Fig. 2.3 Electrical model of a cell membrane. The voltage sources represent the Nernst potentials of the specific ion species, while g_m are the respective ion channel conductances. The current sources represent the Na^+-K^+-pump and are often neglected due to their very small influence on the cell's resting potential. Finally the membrane has a capacitance C_m

Neurons may also have an active mechanism to control the concentration gradient of the chloride ions. If this is not the case the chloride ions redistribute passively only, which means that V_{Cl} is equal to the resting potential of the membrane. There are also large organic anions in the cytoplasm of the cell, denoted as A^-. There are no ion channels for these species, which means they do not contribute to V_m. Finally the membrane is characterized by a membrane capacitance C_m, as depicted in Fig. 2.3.

2.1.3 *Ion Channel Gating*

The membrane is characterized by a resting potential $V_m = V_{rest} \approx -70\,\text{mV}$ that is determined by the values of V_x and $g_{m,x}$ of the model in Fig. 2.3. In this situation $g_{m,x}$ is determined by so-called *rested channels*, which are always open and have a fixed conductance. When V_m becomes more negative (hyperpolarizing) or slightly more positive (depolarizing), the values of $g_{m,x}$ remain approximately constant and the membrane is said to respond in a passive (electrotonic) way.

However, when the membrane potential is depolarized up to a certain threshold, so-called *gated channels* for sodium and potassium that exist in the axon will open; the conductance of these channels depends on the value of V_m. When the membrane voltage is increased to about $-50\,\text{mV}$ (by depolarization), the conductivity of the membrane for sodium ions increases very rapidly. Simultaneously, the potassium ion permeability starts to increase as well, but the process is much slower. This means the sodium ions start to flow from the outside to the inside first, making the inside more positive. When the membrane voltage is increased up to about $20\,\text{mV}$ the potassium conductivity is increased as well and potassium ions begin to flow from the inside to the outside, decreasing the membrane voltage. Finally, the membrane reaches its equilibrium voltage again.

These gated channels are responsible for the regenerative manner in which the action potentials propagate along an axon. If an action potential is generated at some point of the axon, it depolarizes the membrane further down the axon, which will generate a new action potential at that point.

The most widely used model to describe the channel conductance and its rich dynamic properties as a function of the membrane voltage is given by the Hodgkin–Huxley model [3]. Neglecting the Na^+-K^+-pump in Fig. 2.3, the total membrane current i_m is described by:

$$i_m = c_m \frac{dV_m}{dt} + (V_m - V_{Na})g_{m,Na} + (V_m - V_K)g_{m,K} + (V_m - V_{Cl})g_{m,Cl} \tag{2.2}$$

The conductances are given by the following equations:

$$g_K = G_K n^4 \tag{2.3a}$$

$$g_{Na} = G_{Na} m^3 h \tag{2.3b}$$

$$g_{Cl} = G_L \tag{2.3c}$$

The conductances G_{Na}, G_K, and G_L are constants, while the factors n, m, and h are described by the following differential equations:

$$\frac{dm}{dt} = \alpha_m(1 - m) - \beta_m m \tag{2.4a}$$

$$\frac{dh}{dt} = \alpha_h(1 - h) - \beta_h h \tag{2.4b}$$

$$\frac{dn}{dt} = \alpha_n(1 - n) - \beta_n n \tag{2.4c}$$

Here the factors α_x and β_x depend on the membrane overpotential $V' = V_m - V_{rest}$ via:

$$\alpha_m = \frac{0.1 \cdot (25 - V')}{\exp \frac{25 - V'}{10} - 1} \quad \beta_m = \frac{4}{\exp \frac{V'}{18}} \tag{2.5a}$$

$$\alpha_h = \frac{0.07}{\exp \frac{V'}{20}} \quad \beta_h = \frac{1}{\exp \frac{30 - V'}{10} + 1} \tag{2.5b}$$

$$\alpha_n = \frac{0.01(10 - V')}{\exp \frac{10 - V'}{10} - 1} \quad \beta_n = \frac{0.125}{\exp \frac{V'}{80}} \tag{2.5c}$$

These equations model the electrical response of a membrane containing gated ion channels. The equations will be used to obtain the response of the membrane during electrical stimulation, described in the next section.

2.2 Stimulation of Neural Tissue

As shown in the previous section, neurons are electrochemical systems. Drugs have long been used to alter the chemical component of this system for therapeutic purposes. Drawback of this approach is that drugs generally affect the whole nervous system and are therefore easy to induce unwanted side effects.

Neural stimulation uses the electric component of the nervous system by locally altering the membrane voltage of the neurons. Neural stimulation has been applied using electrical, magnetic, and more recently, optical and acoustic stimuli.

Electrical stimulation of neural tissue uses an electric field to artificially recruit neurons in order to tap into the neural system on a functional level. Using stimulation neurons can be forced to generate action potentials (activation) or prevented from generating them (inhibition). This section reviews the stimulation process from the stimulator up to the membrane voltage.

Stimulation is considered at three different levels:

- The electrode level: the electrodes and the tissue are modeled using an equivalent electrical circuit and form the load of a stimulator.
- The tissue level: the tissue itself is modeled as a volume conductor and the electrical stimulation leads to an electric field inside this volume conductor.
- At the neuronal level: the electric field will influence the local extracellular membrane potential of a neuron, which will ultimately trigger or suppress an action potential.

2.2.1 *The Electrode Level: Electrode–Tissue Model*

The electrical energy that is used to recruit neurons is injected into the target area by means of electrodes. These electrodes form the load of the stimulator circuit and in order to design an efficient and safe system it is essential to have an electrical model of this load.

In the electrode electrons are the charge carrying particles, while in tissue charge is carried by ions. This means that the system needs to be considered as a electrochemical system. The equivalent circuit is typically divided into two parts: the electrode–tissue interface Z_{if} and the tissue impedance Z_{tis} as is shown in Fig. 2.4.

2.2.1.1 Electrode–Tissue Interface

Processes at the Electrode–Tissue Interface

At the interface of the electrode and the tissue some interaction must take place between the electrons in the electrodes and the ions in the tissue. There are two types of interactions possible: charge accumulation and electrochemical reactions.

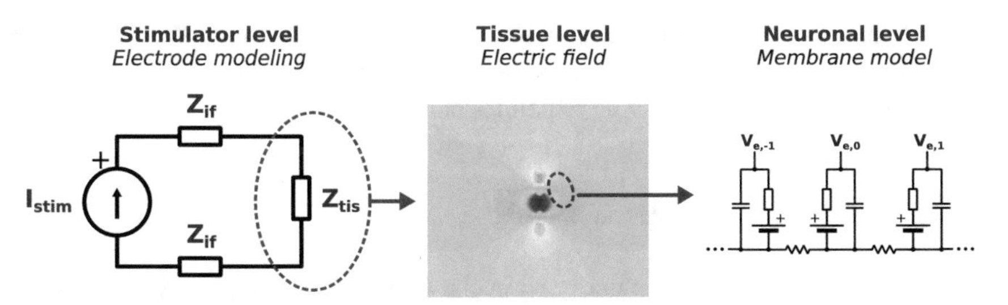

Fig. 2.4 Three levels of hierarchy at which neural stimulation is considered. The stimulator level considers the electrical equivalent circuit that is connected to the stimulator circuit. The tissue level zooms in on the tissue impedance by considering it as a volume conductor. The neuronal levels zoom further in on the neuron itself and how the membrane voltage develops during stimulation

Charge accumulation is simply the accumulation of charge carriers (double layer charging) near the interface. In this mechanism no charge is transferred between the electrode and the tissue. In case of *electrochemical reactions* charge is transferred at the interface by means of redox reactions. One electrode acts as the anode: an oxidation half-reaction describes how the electrode loses electrons. The other electrode acts as the cathode: a reduction half-reaction shows how the electrode accepts electrons.

Each half-reaction is first of all described by an electrode potential V_n that resembles the built-in potential of the electrode in equilibrium that results from the difference in electrochemical potential between the electrode and the tissue. Like Eq. (2.1) for the cell membrane, this potential is described using the Nernst equation:

$$V_n = V_0 - \frac{RT}{zF} \ln \frac{a_{red}}{a_{ox}} \tag{2.6}$$

Here V_0 is the half-cell potential under standard conditions, while the second term corrects for deviations from these conditions. The factors a_{red} and a_{ox} are the activity of the reduction and oxidation species in the half cell, which are closely related to the concentrations of the gaseous and aqueous species and z is the valence of the reaction.

When an overpotential $\eta_a = V_{if} - V_n$ is established over the interface, the kinetics of the reactions are described using the electrochemical current density i_{net} in the Butler–Volmer equation [4]:

$$i_{net} = i_0 \left\{ \frac{[O]_{0,t}}{[O]_\infty} \exp\left(-\alpha_c z f \eta_a\right) - \frac{[R]_{0,t}}{[R]_\infty} \exp\left((1-\alpha_c) z f \eta_a\right) \right\} \tag{2.7}$$

Here i_0 is the exchange current density, α_c is the cathodic transfer coefficient, $f \equiv F/RT$, and $[O]_{0,t}/[O]_\infty$ and $[R]_{0,t}/[R]_\infty$ are the ratios between concentrations of the oxidized and reduced species at the electrode and in the rest of the tissue, respectively. For high overpotentials the ratios $[O]_{0,t}/[O]_\infty$ and $[R]_{0,t}/[R]_\infty$ will decrease, causing the reaction to become mass limited at the limiting current $i_{L,a}$ or $i_{L,c}$ for the anodic and cathodic reactions, respectively.

Modeling of the Electrode–Tissue Interface

The two different processes mentioned in the previous paragraph now need to be translated to equivalent electrical models. The interface is characterized by two types of mechanisms: reversible and irreversible processes [5].

Reversible currents are characterized by the fact that they store charge at the interface. This can be due to charge accumulation (capacitive currents), but also due to reversible electrochemical reactions (pseudo-capacitive currents). In the case of, for example, hydrogen plating, hydrogen atoms are bound to the electrode surface, effectively storing charge at the interface. Both processes are characterized by their reversibility: the stored charge can be recovered by reversing the electrical current. The reversible currents are modeled with a capacitor C_{dl}.

Irreversible currents, also often labeled as the faradaic currents, are due to irreversible processes such as electrochemical processes in which the reaction products cannot be reversed, such as oxygen evolution. They are modeled with a charge transfer resistor R_{CT}. Furthermore it is often chosen to model the equivalent built-in potential V_{eq} (combining the relative contribution of V_n for each electrochemical reaction) as a voltage source in series with R_{CT}.

The values of C_{dl} and R_{CT} highly depend on the size, geometry, and materials used for the electrodes. Furthermore, the complex kinetics of the electrochemical reactions according to Eq. (2.7) make both components highly non-linear. In many applications the model is linearized.

Note that the naming of C_{dl} (dual layer capacitor) and R_{CT} (charge transfer resistor) can be confusing. A dual layer capacitor suggests that it only models charge accumulation, while it can also model the charge transfer of pseudo-capacitive current (which is *not* modeled in the charge transfer resistance!).

Electrodes with almost exclusively reversible currents (e.g., platinum electrodes) are called polarizable electrodes: by injecting current through these electrodes, the interface will be polarized. Electrodes with predominantly faradaic currents are non-polarizable electrodes (e.g., Ag/AgCl electrodes).

2.2.1.2 Tissue Impedance

The tissue impedance is the electrical equivalent circuit of the tissue itself. The voltage over this component is directly related to the strength of the electric field, as will be seen in the next subsection. The second order spatial difference of this electric field is subsequently important for recruiting the neurons. Current based stimulation is often preferred to make the voltage over the tissue (and hence the electric field strength) independent of the interface impedance Z_{if}: $V_{tis} = I_{stim} Z_{tis}$.

The impedance of the tissue first of all heavily depends on the electrode geometry: if the electrodes are big (i.e., have a large effective area), the impedance will be low. Because of their low impedance, big electrodes need a higher current than small electrodes to create the same voltage and thus electric field strength. However these big electrodes create a much larger electric field, therefore affecting a large population of neurons. Small electrodes create a much smaller electric field, affecting much less neurons while using only a little bit of current.

It depends on the application whether small or big electrodes are used. That is why such a wide range of currents and impedances is encountered in literature: from a few μA for micro-electrodes with Z_{tis} in the order of 100 kΩ to tens of mA for big electrodes with R_{tis} in the order of 10 Ω.

Besides electrode size the impedance also depends on the tissue properties: it can incorporate as many of the tissue properties as desired: non-linearity, anisotropy, inhomogeneous, time-variant, and dynamic properties. However many of these properties make it hard to handle the model in standard circuit simulators. In many cases the tissue is simply modeled using a resistor R_{tis} only. If the dynamic properties are important a capacitor C_{tis} can be placed in parallel as a first approximation.

In literature or in electrode specifications one often finds a single value for the "electrode impedance." Usually this value corresponds to the impedance measured at 1 kHz with very low excitation levels. This value represents the equivalent impedance of the whole system depicted in Fig. 2.5.

2.2.2 Tissue Level: Electric Field Distribution

In the previous subsection the tissue was considered as a discrete impedance Z_{tis} as part of the load of the stimulator. In this section we zoom in on the tissue itself by considering it as a volume conductor. Neural tissue has non-linear, inhomogeneous, anisotropic, and time-variant electrical properties.

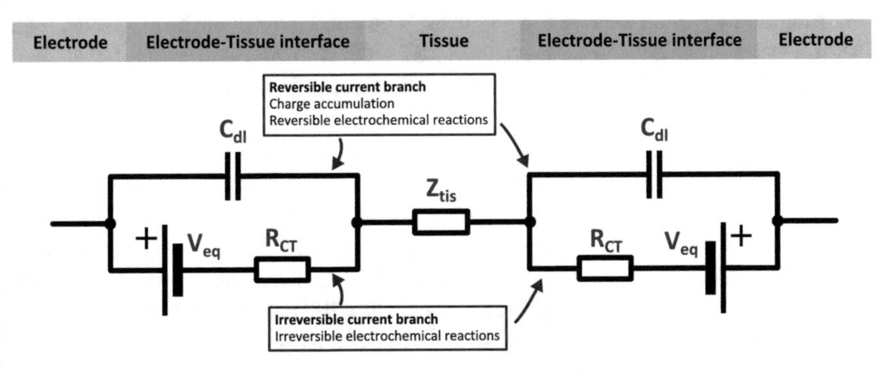

Fig. 2.5 Equivalent electrical model of the electrode system. The interface model consists of a capacitive branch (C_{dl}) and a faradaic branch (R_{CT} and V_{eq}). The tissue itself is modeled with impedance Z_{tis}. Note that all components have non-linear properties

The electric field $\overline{E}$ is in general found using Gauss' law:

$$\nabla \cdot \overline{E} = \frac{\rho}{\epsilon} \tag{2.8}$$

Here ρ is the charge density and ϵ is the permittivity. Assuming quasi-static conditions, the curl of $\overline{E}$ must be zero according to Faraday's law of induction:

$$\nabla \times \overline{E} = -\frac{d\overline{B}}{dt} = 0 \tag{2.9}$$

According to the Helmholtz decomposition theorem, each continuously differentiable vector can be decomposed in a curl-free and a divergence free component. The curl being zero (Eq. (2.9)), the potential is related to the electric field via:

$$\overline{E} = -\nabla \Phi \tag{2.10}$$

Substituting Eq. (2.10) into Eq. (2.8) gives the Poisson equation:

$$-\nabla^2 \Phi = \frac{\rho}{\epsilon} \tag{2.11}$$

For tissue it holds in general that $\rho = 0$, which reduces Poisson equation into the Laplace equation: $\nabla^2 \Phi = 0$. The goal is now to solve this equation for certain boundaries as set by the tissue and the stimulation electrodes.

A way to numerically solve the Laplace equation is by using finite element method analysis. A model can be constructed for the electrodes as well as for the tissue. For illustrative purposes a model is created for the electrode leads in Fig. 2.6a. These electrode leads are made of silicone rubber and can be used for spinal cord stimulation. The platinum electrodes are ring shaped with a diameter of 1.5 mm and a height of 3 mm.

A 2D model of these electrodes was constructed in the ANSYS software environment, which is depicted in Fig. 2.6a. The conductivity of the electrodes and the insulation was chosen to be $\sigma_s = 9.52 \cdot 10^6$ S/m and $\sigma_p = 2 \cdot 10^{-14}$ S/m, respectively. The tissue was modeled as an isotropic and homogeneous plane with a conductivity of $\sigma_t = 0.3$ S/m [6].

In Fig. 2.6 the simulation results of the potential distribution for various stimulation strategies are depicted. In Fig. 2.6b only one electrode is driven with a current source of 10 mA (monopolar operation), while the edges of the tissue plane were connected to 0 V to mimic a large counter electrode. In Fig. 2.6c one electrode is used as the cathode, while the other is the anode (bipolar stimulation). In Fig. 2.6d one electrode is used as a cathode and is driven with 10 mA, while the other two electrodes both act as the anode and are driven with 5 mA each.

As can be seen from Fig. 2.6, the shape and size of the electric field heavily depend on the choice of the electrodes, as well as on how they are operated.

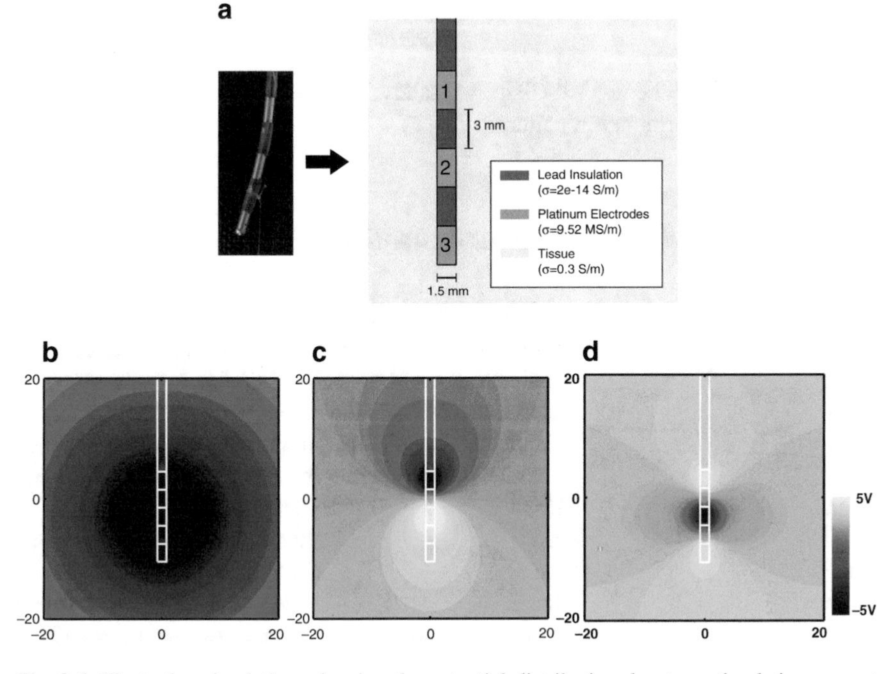

Fig. 2.6 Illustrative simulations showing the potential distribution due to a stimulation current through electrodes. In (**a**) the finite element model is sketched, while in (**b**), (**c**), and (**d**) the potential distributions of monopolar, bipolar, and tripolar stimulation configurations are shown, respectively. The dimensions on the axis are in mm

2.2.2.1 Monopole Example

For specific cases the electric field can be solved analytically. Here we will analyze the potential distribution of a point source (monopole) in an infinite homogeneous isotropic volume conductor with conductivity σ. This situation can approximate a real situation when the distance from the electrode becomes large with respect to the electrode size.

The current from the source will spread out in the volume conductor equally to all directions, hence the current density at a distance r from the source decreases proportionally with the area of a sphere:

$$\overline{J} = \frac{I_{stim}}{4\pi r^2}\overline{a}_r \tag{2.12}$$

Here I_{stim} is the stimulation current and $\overline{a}_r$ is the unit vector in the radial direction with the point source at the origin. Using Ohm's law $\overline{J} = \sigma\overline{E}$ and realizing that the electric field only varies over the radial direction, the potential is found using Eq. (2.10) via integration of the electric field over r as:

$$\Phi = \frac{I_{stim}}{4\pi\sigma r} \tag{2.13}$$

Equation (2.13) will be used in Chap. 4 to analyze the electric field resulting from a particular type of stimulation pattern.

2.2.3 Neuronal Level: Axonal Activation

Knowing the potential distribution in the tissue, it is possible to analyze the resulting membrane potential in the neurons. Although activation or inhibition of the neurons can occur anywhere, often axonal activation is considered [7–9]. A cable model is used for which the axon is divided into segments that contain the membrane model of Fig. 2.3: the membrane capacitance C_m, the resting potential V_{rest}, and a resistance Z_{HH} modeling the membrane channels. Each membrane model is connected using an intracellular resistance R_i. This is schematically depicted in Fig. 2.7.

The cable model is now placed in the potential field that was found in the previous section, which will determine the extracellular potentials $V_{e,n}$ along the axon. Based on these potentials, the membrane voltage $V_{m,n} = V_{i,n} - V_{e,n}$ at node n can be found by solving the following equation that follows directly from Kirchhoff's laws [7]:

$$\frac{dV_{m,n}}{dt} = \frac{1}{C_m}\left[\frac{1}{R_i}(V_{m,n-1}-2V_{m,n}+V_{m,n+1}+V_{e,n-1}-2V_{e,n}+V_{e,n+1})-i_{HH}\right] \tag{2.14}$$

Here i_{HH} is the current described by the Hodgkin–Huxley equations through the impedance Z_{HH}. As will be shown in Chap. 4, this equation can be used to analyze the membrane voltage during a stimulation pulse. If the membrane voltage is elevated above the threshold for a certain amount of time, the dynamics of the Hodgkin–Huxley equations predict that an action potential will be generated.

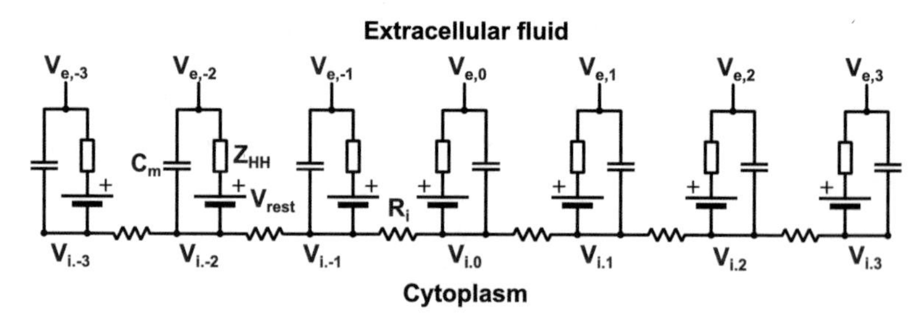

Fig. 2.7 Axon model used to evaluate the membrane voltage $V_m = V_i - V_e$ during a stimulation. The extracellular potentials $V_{e,i}$ are determined by the electric field resulting from the stimulation current

From Eq. (2.14) the source term due to the electric field can be isolated and is found as:

$$f_n = \frac{V_{e,n-1} - 2V_{e,n} + V_{e,n+1}}{\Delta x^2} \tag{2.15}$$

Here it is used that $R_i = 4\rho_i \Delta x/(\pi d^2)$ and $C_m = \pi d L c_m$ in which ρ_i is the intracellular resistivity, d is the axon diameter, L is the length of membrane segment, c_m is the membrane capacitance per unit area, and Δx is the segmentation length of the membrane. f_n is called the activation function [9] and for $\Delta x \to 0$, it becomes the second order derivative of the extracellular potential. As a first approximation it can be stated that as long as $f_n > 0$ for a certain axon segment, V_m will increase and hence it is likely that activation will occur. This assumes that the stimulation is strong enough and enabled long enough to allow V_m to rise above the threshold.

Thus, in conclusion, for a membrane to be recruited, f_n needs to be sufficiently positive for a certain amount of time. This property can be translated to a relation between the minimum required stimulation current I_{stim} and the stimulation duration t_{stim}. This property is well known and is summarized in the *charge–duration curve*, which shows the following hyperbolic relationship: $I_{stim} = a/t_{stim} + b$ [10]. This curve is sketched in Fig. 2.8. The minimum stimulation current $I_{stim} = b$ required to achieve stimulation is called the rheobase [10], while the $t_{stim} = a/b$ corresponding with twice the rheobase is called the chronaxie. The chronaxie is the point at which the energy of the stimulation pulse E_{pulse} is minimal:

$$E_{pulse} = I_{stim}^2 Z_{load} t_{stim} = \left(\frac{a^2}{t_{stim}} + b^2 t_{stim}\right) Z_{load} \tag{2.16}$$

Here Z_{load} is the impedance of the electrodes. The minimum energy is found to be equal to the chronaxie via:

$$\frac{dE_{pulse}}{dt_{stim}} = \left(\frac{-a^2}{t_{stim}^2} + b^2\right) Z_{load} = 0 \;\rightarrow\; t_{stim} = \frac{a}{b} \tag{2.17}$$

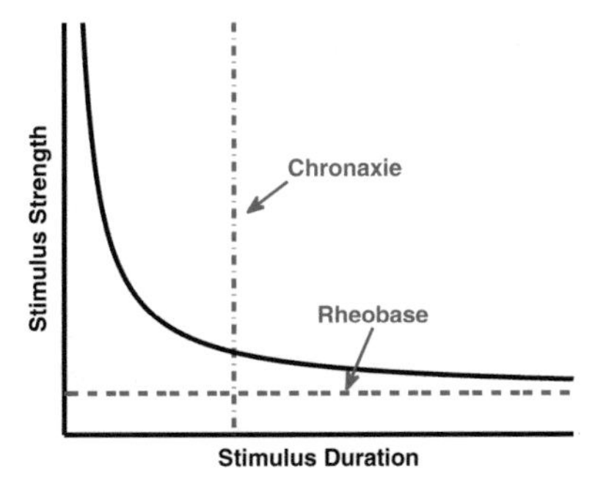

Fig. 2.8 Strength–duration curve showing the required stimulation intensity as a function of the stimulus duration. The rheobase and chronaxie are depicted as well

2.3 Conclusions

In this chapter an overview has been given of the electrophysiological principles of neurons. It was shown how the electrochemical properties of the neural cell membrane are responsible for the generation of action potentials. Furthermore, it was shown how electrical stimulation influences the neural tissue: a stimulation current will generate an electric field in the tissue, which will influence the membrane voltage of the neurons by depolarizing or hyperpolarizing the membrane. The minimum required stimulation intensity is described by the strength–duration curve, showing a hyperbolic relationship between the stimulation amplitude and duration.

References

1. Kandel, E., Schwartz, J.H., Jessell, T.: Principles of Neural Science. McGraw-Hill, New York (2000)
2. Malmivuo, J., Plonsey, R.: Bioelectromagnetism – Principles and Applications of Bioelectric and Biomagnetic Fields. Oxford University Press, New York (1995)
3. Hodgkin, A.L., Huxley, A.F.: A quantitative description of membrane current and its application to conduction and excitation in nerve. J. Physiol. **117**(4), 500–544 (1952)
4. Zoski, C.: Handbook of Electrochemistry. Elsevier, Amsterdam (2006)
5. Merrill, D.R., Bikson, M., Jefferys, J.G.R.: Electrical stimulation of excitable tissue – design of efficacious and safe protocols. J. Neurosci. Methods **141**, 171–198 (2005)
6. Butson, C.R., McIntyre, C.C.: Tissue and electrode capacitance reduce neural activation volumes during deep brain stimulation. Clin. Neurophysiol. **116**, 2490–2500 (2005)
7. McNeal, D.R.: Analysis of a model for excitation of myelinated nerve. IEEE Trans. Biomed. Eng. **23**(4), 329–337 (1976)
8. Warman, E.N., Grill, W.M., Durand, D.: Modeling the effects of electric fields on nerve fibers: determination of excitation thresholds. IEEE Trans. Biomed. Eng. **39**(12), 1244–1254 (1992)
9. Rattay, F.: Analysis of models for extracellular fiber stimulation. IEEE Trans. Biomed. Eng. **36**(7), 676–682 (1989)
10. Lapicque, L.: Définition Expérimentale de l'excitabilité. C. R. Séances Soc. Biol. Fil. **61**, 280–283 (1909)

Chapter 3
Electrode–Tissue Interface During a Stimulation Cycle

Abstract While neural stimulation has shown to be a very effective treatment, it is important to ensure safe operation to protect both the neural tissue as well as the electrodes. This chapter starts by shortly reviewing the potential damage mechanisms that can occur when neural stimulation is used. One of these mechanisms, the irreversible charge transfer processes at the electrode–tissue interface, is considered in the following two sections in more detail. First the consequences of using coupling capacitors are analyzed. Coupling capacitors are often claimed to improve the charge cancellation and therefore protect the electrode–tissue interface. In Sect. 3.2 it is verified whether this is indeed the case. In the last section a novel stimulation technique is introduced that aims to return the electrode–tissue interface to equilibrium after a stimulation cycle.

3.1 Damage Mechanisms

Implanting a stimulator can result in tissue damage. From some kinds of damage the brain tissue is able to recover, for other kinds of damage this is not possible and the damage is irreversible. Damage can be divided in two main categories [1]: mechanically and electrically induced.

3.1.1 Mechanically Induced Damage

Mechanical damage is caused by the invasive placement of the electrodes and/or the stimulator. Upon implantation the body will accumulate connective tissue around the electrodes to encapsulate them. In peripheral nerves this might lead to thickening in the connective tissue in the epineurium, subperineurial and endoneurial[1] tissue. In severe cases endoneurial edema, demyelination, and axonal degeneration are

[1]In peripheral nerves, the individual nerve fibers are surrounded by a sheath called the *endoneurium*. The nerve fibers are grouped together in *fascicles*, which are surrounded by a sheath called the *perineurium*. The fascicles group together to form a nerve, which is surrounded by the *epineurium*.

M. van Dongen, W. Serdijn, *Design of Efficient and Safe Neural Stimulators*, Analog Circuits and Signal Processing, DOI 10.1007/978-3-319-28131-5_3

reported [1, 2]. In the central nervous system neuronal loss is observed due to the use of penetrating micro-electrodes in the cerebral cortex [3].

Mechanical damage can be minimized by careful design of the electrodes and implantation procedures. Examples include the use of bio-compatible materials and the avoidance of situations in which the electrodes and/or cables cause stress, compression, or abrasion (relative movement between the electrodes and the nerves).

3.1.2 Electrically Induced Damage

The mechanisms underlying electrically induced damage are not fully understood, but a few aspects have been hypothesized to attribute to neuronal damage.

- *Hyper-activation*
  Neuronal damage is found to be related with the activation of the axons: neuronal damage is increased in axons that are activated more frequently and/or stronger. Similarly, recruiting axons less frequently will lead to less damage [2]. Therefore it is hypothesized that neuronal damage results from the hyper-activation of neurons [4]: too much activation might cause an imbalance in the concentrations of the various ions in- and outside the neurons (Fig. 2.2).
- *Electroporation*
  Electroporation is the sudden change of the cell membrane conductivity due to the application of a strong electric field [5]. This conductivity change is not due to the opening of ion channels, but due to the formation of pores in the membrane. Electroporation is hypothesized to be responsible for neuronal damage by examining the current density thresholds for neuronal damage in relation with the stimulation duration [6].
- *Electrochemical damage*
  In Sect. 2.2.1 it was shown that the electrode–tissue interface has an electrochemical nature. Therefore, during stimulation it has to be assured that no harmful electrochemical reactions are triggered that can cause damage to the electrodes and/or the tissue. Ideally all charge transfer processes at the interface should be reversible.

The first and second categories are prevented by choosing sufficiently low stimulation parameters. The third category requires the electrode–tissue interface to operate within safe electrochemical boundaries. These boundaries are also determined by the stimulation parameters, but they require additional safety mechanisms from the stimulator circuit. This chapter discusses two circuit techniques that aim to improve or investigate the electrochemical operating conditions of the electrodes.

For polarizable electrodes the electrochemical safety is usually ensured by considering the voltage over the capacitive electrode–tissue interface C_{dl} (Fig. 2.5) of the electrodes: this voltage should not exceed a certain window to avoid irreversible charge transfer processes [7]. This means first of all that the total amount of charge that can be injected during a stimulation pulse is limited (limiting the stimulation parameters to the reversible charge injection limits [8]). Second of all, by using a biphasic stimulation scheme, charge accumulation over multiple stimulation cycles on C_{dl} is avoided [9]. Section 3.2 zooms in on this aspect by analyzing the consequences of using coupling capacitors. Section 3.3 introduces a stimulation circuit that aims to return the electrode–tissue interface back to equilibrium.

3.2 The Consequences of Using Coupling Capacitors

The use of coupling capacitors between the stimulator and the electrodes is widely considered to be an effective safety mechanism [10, 11]. Various advantages towards the use of coupling capacitors have been identified [12]. The first important advantage is the prevention of DC currents in the event of device failure. If, for example, one of the electrodes shorts to the supply voltage, the coupling capacitor will prevent a prolonged DC current through the electrodes.

The second important advantage that is attributed to coupling capacitors is that they improve the performance of passive charge balancing techniques [13]. Charge balancing is important for polarizable electrodes to keep the electrode–tissue interface within an electrochemically safe regime [14]. In practice this means that the interface voltage should return to zero after a stimulation cycle. A coupling capacitor helps due to its high-pass characteristics, which limits the flow of DC currents and hence no net charge can be injected into the tissue.

A disadvantage of coupling capacitors is that their required value is often too high to be integrated on an IC [13] and hence they are realized using bulky external components. Many studies have focused on designing stimulator output stages with accurate charge balancing circuits [15, 16] in order to eliminate the need of coupling capacitors. Others have proposed high frequency operation to reduce their size [17]. Indeed the results seem to suggest that the proposed mechanisms are good enough to prevent charge accumulation on the tissue even without coupling capacitors. However, it is not clear how these systems can guarantee safety in the event of a device failure. For this reason many stimulator systems still require the use of coupling capacitors.

Although widely used, it often seems overlooked that a coupling capacitor eliminates control over the DC voltage across the electrodes. As will be shown [18], it is therefore possible for an offset voltage V_{os} to develop over the electrode–tissue interface, even when the electrodes and capacitors are shorted in between the stimulation pulses and charge balanced biphasic stimulation is used. If V_{os} becomes too large, the electrode–tissue interface may leave the electrochemically safe regime, triggering the production of potentially dangerous reaction products. In this case the intended safety mechanisms of the coupling capacitor create the opposite result: a potentially dangerous situation is created. In this section the value of V_{os} is analyzed over various operating conditions, both analytically as well as experimentally. This gives insight in when V_{os} is exceeding a predefined safe regime.

3.2.1 Methods

A basic setup of a biphasic stimulator system is depicted in Fig. 3.1a: the coupling capacitor C_c is connected in series with the stimulator and the electrodes. The stimulation source in Fig. 3.1a is a biphasic constant current stimulator with a cathodic first stimulation pulse with amplitude I_c and duration t_c. The anodic charge cancellation phase follows with amplitude I_a and duration t_a. Most stimulator systems apply a passive charge balancing scheme, in which the series connection of the electrodes and coupling capacitor are shorted after the stimulation cycle by closing switch S_1 to discharge C_{dl}. The duration of shortening t_{dis} is determined by the repetition rate $f_{stim} = 1/t_{stim}$ of the stimulation, since S_1 needs to be opened again when the next stimulation cycle starts.

As shown in Fig. 3.1a, the electrodes are modeled as a resistance R_s in series with capacitor C_{dl} and resistor R_{ct} that model the electrode–tissue interface, as discussed previously in Chap. 2 [19]. The electrodes used in this study are single percutaneous octrode leads (manufactured by ANS, currently St. Jude Medical): they consist of eight ring shaped platinum contacts that are distributed on a single lead. Each electrode has a diameter of 1.5 mm and a width of 3 mm (area 0.14 cm^2). A picture of the stimulation setup and the electrodes is depicted in Fig. 3.1b. The electrodes were

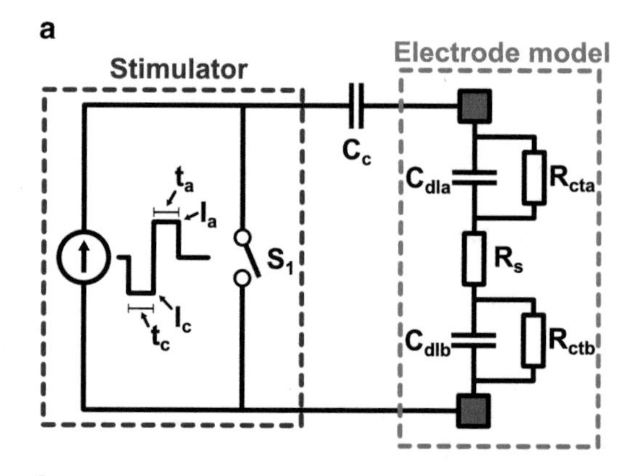

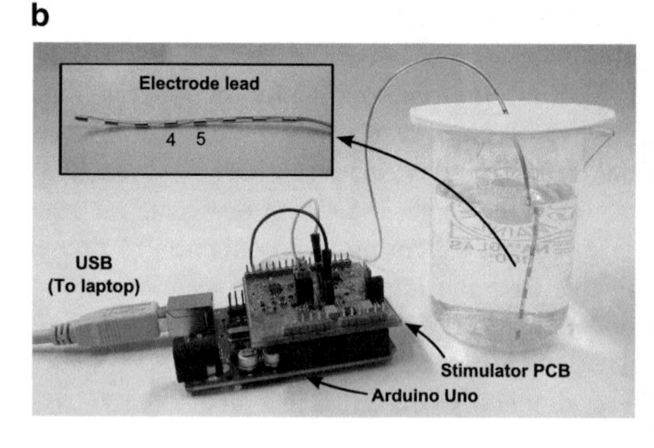

Fig. 3.1 In (**a**) a basic setup of a biphasic constant current stimulator system is shown that includes a coupling capacitor C_c and an electrode model. In (**b**) a picture of the measurement setup is shown with a detail of the electrode lead where contacts 4 and 5, which were used for stimulation, are indicated

submerged in a phosphate buffered saline (PBS) solution containing the following: 1.059 mM KH_2PO_4, 155.172 mM NaCl, and 2.966 mM Na_2HPO_4-$7H_2O$ (pH 7.4, Gibco® Life Technologies™). The electrodes were connected in a bipolar fashion by selecting contacts 4 and 5 as the anode and cathode. The other contacts were left floating.

Using an HP4194A impedance analyzer (excitation amplitude 0.1 V), it was found that for these electrodes in a saline solution $R_s \approx 100\,\Omega$ and $C_{dl} \approx 1.5\,\mu$F. Here C_{dl} is the capacitive part of both electrode–tissue interfaces combined. The value of $R_{ct} = 1\,\text{M}\Omega$ (also combining both interfaces) was determined by measuring the voltage over the electrodes due to a 5 nA DC current from a Keithley 6430 sub-femtoamp sourcemeter. These type of electrodes are typically used for spinal cord stimulation and the stimulation amplitudes used in this chapter are based on the specifications of the EON™ IPG (also from St. Jude Medical) [20].

3.2.1.1 Determining V_{os}

After the anodic phase both C_c and C_{dl} will be charged. Upon closing S_1 these capacitors will be discharged with a time constant

$$\tau_{dis} = R_s C_{eq} \qquad C_{eq} = \frac{C_c C_{dl}}{C_c + C_{dl}} \tag{3.1}$$

If S_1 would be closed sufficiently long, a pseudo steady-state is reached in which:

$$V_{Cc} + V_{Cdl} = 0 \tag{3.2}$$

Here V_{Cc} is the voltage over C_c. If S_1 is closed even longer, C_{dl} will continue to discharge through R_{ct} with time constant $\tau_2 = R_{ct} C_{dl}$ until $V_{Cdl} = 0$ V and the actual steady-state is reached. However, usually $t_{dis} \ll \tau_2$ and therefore only the pseudo steady-state is reached.

Note that Eq. (3.2) does *not* guarantee that $V_{Cdl} = 0$ in pseudo steady-state: it is an under-determined equation and $V_{Cc} = -V_{Cdl}$ can have any value. Only when both C_c and C_{dl} are ideal capacitors, the same current is flowing through both capacitors during a stimulation cycle, which causes $V_{Cc} = V_{Cdl} = 0$ V in pseudo steady-state. If these requirements are not met (e.g., when $R_{ct} \neq \infty$), the current though C_c does not equal to the current through C_{dl}, which will cause $V_{Cdl} = -V_{Cc} \neq 0$ in pseudo steady-state. This charge imbalance can accumulate over many stimulation cycles, which creates an offset in V_{os} over V_{Cdl}.

We refer to Fig. 3.2 to analyze V_{Cdl} when after many stimulation cycles the offset voltage V_{os} is stable. In order for this voltage to be stable, the average current through R_{ct} must be zero such that no charge is lost that causes an inequality in the charge accumulated on C_{dl} with respect to C_c. Therefore it must hold that the average value of V_{Cdl} (and hence the area as indicated in Fig. 3.2) must be zero as well.

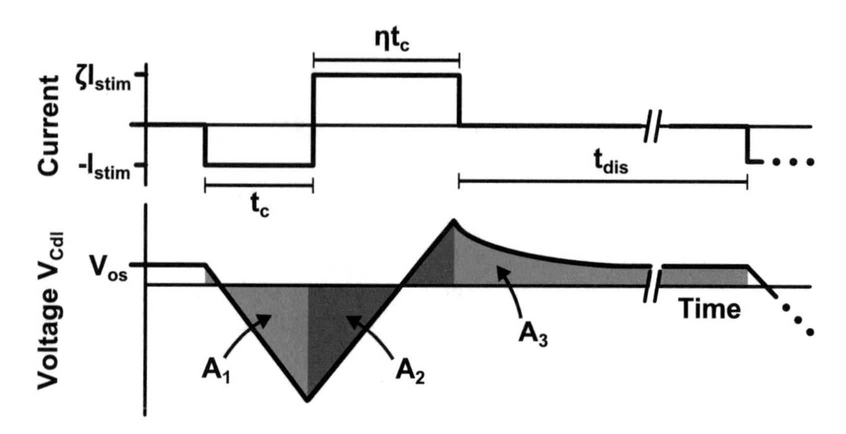

Fig. 3.2 Schematic plot of V_{Cdl} during a biphasic stimulation cycle with charge mismatch. When V_{os} is stable the area $A_1 + A_2 + A_3$ equals zero

To find the value of V_{os} for which this requirement is met, it is assumed that the cathodic stimulation phase is characterized by a duration t_c and amplitude $I_c = I_{stim}$. In the anodic phase both the duration $t_a = \eta t_c$ and the amplitude $I_a = \zeta I_{stim}$ can include mismatch. Furthermore it is assumed that $\tau_{dis} \ll t_{dis}$ (such that pseudo steady-state is reached) and that R_{ct} is large enough to be neglected in the analysis (but as stated above it must be finite). The areas A_1, A_2, and A_3 are found as:

$$A_1 = \int_0^{t_c} \left(V_{os} - \frac{I_{stim} t}{C_{dl}} \right) dt = V_{os} t_c - \frac{I_{stim} t_c^2}{2C_{dl}} \tag{3.3a}$$

$$A_2 = \int_0^{\eta t_c} \left(V_{os} - \frac{I_{stim} t_c}{C_{dl}} + \frac{\zeta I_{stim} t}{C_{dl}} \right) dt = V_{os} \eta t_c - \frac{I_{stim} \eta t_c^2}{C_{dl}} + \frac{\zeta I_{stim} (\eta t_c)^2}{2C_{dl}} \tag{3.3b}$$

$$A_3 = \int_0^{t_{dis}} \left(V_{os} - (1 - \zeta\eta) \frac{I_{stim} t_c}{C_{dl}} \exp\left(\frac{-t}{R_s C_{dl}} \right) \right) dt = V_{os} t_{dis} - (1 - \zeta\eta) I_{stim} t_c R_s \tag{3.3c}$$

By setting $A_1 + A_2 + A_3 = 0$ and solving for V_{os}, the following equation is obtained:

$$V_{os} = \frac{(0.5 + \eta - 0.5\zeta\eta^2) I_{stim} t_c^2 + (1 - \zeta\eta) I_{stim} C_{dl} t_c}{C_{dl} t_c (1 + \eta) + t_{dis}} \tag{3.4}$$

If $\zeta = \eta = 1$, which means that perfectly charge balanced stimulation is applied, the following equation holds:

$$V_{os} = \frac{I_{stim} t_c^2}{C_{dl}(2t_c + t_{dis})} = \frac{I_{stim} t_c^2}{C_{dl} t_{stim}} \tag{3.5}$$

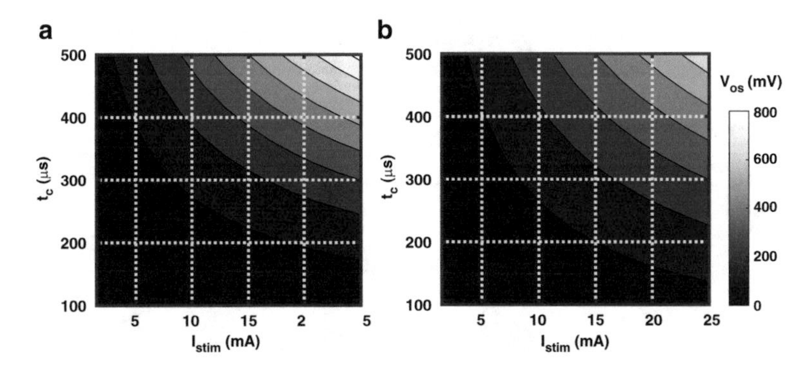

Fig. 3.3 Overview of the pseudo steady-state offset voltages V_{os} for a variety of stimulation settings. In (**a**) a perfectly charge balanced stimulation waveform is chosen and V_{os} is determined according to Eq. (3.5) with $f_{stim} = 200$ Hz. In (**b**) a monophasic stimulation waveform is used and V_{os} is determined using Eq. (3.4) with $\eta = 0$

In Fig. 3.3a the value of V_{os} is depicted for a charge balanced stimulation cycle with $f_{stim} = 200$ Hz that includes a coupling capacitor. For small I_{stim} and t_c the value of V_{os} is small and will have negligible influence on the system. However for larger stimulation intensities V_{os} starts to increase towards several hundreds of millivolts (up to 800 mV for the maximum intensity).

Equation (3.4) can also be used to analyze monophasic stimulation patterns by choosing $\eta = 0$. In Fig. 3.3b the values of V_{os} are plotted for this situation. Somewhat surprisingly V_{os} is smaller than the biphasic charge balanced stimulation. This can be explained by the fact that due to the relatively low value of R_s the discharge current during t_{dis} is larger than I_{stim} and hence the electrodes discharge faster towards pseudo steady-state as compared to the biphasic stimulation waveform.

3.2.1.2 Verifying V_{os}

To verify Eq. (3.4), the response of an electrode system was analyzed using both simulations as well as measurements in a saline bath. To simulate the response of these electrodes, the circuit from Fig. 3.1 was implemented in a simulator (LT-Spice). Switch S_1 was chosen to have $R_{off} = 10\,\mathrm{M\Omega}$ to mimic the limited output impedance of the current source and $R_{on} = 10\,\Omega$. The stimulation current was chosen to be $I_{stim} = 1.5$ mA ($\zeta = 1$), while an 8 % charge mismatch was introduced by making $t_c = 460\,\mu$s and $t_a = 500\,\mu$s ($\eta = 1.087$). After the stimulation cycle, switch S_1 was closed for $t_{dis} = 9$ ms before the next stimulation pulse is started. This makes the stimulation repetition rate slightly higher than 100 Hz.

Using Eq. (3.1) it is found that $t_{dis} > 60\tau_{dis}$, which means that V_{Cdl} and V_{Cc} can be assumed to have reached their pseudo steady-state values. Also $\tau_2 = 1.5\,\mathrm{s} \ll t_{dis}$, which means that the system will stay in pseudo steady-state and will not have the opportunity to fully discharge.

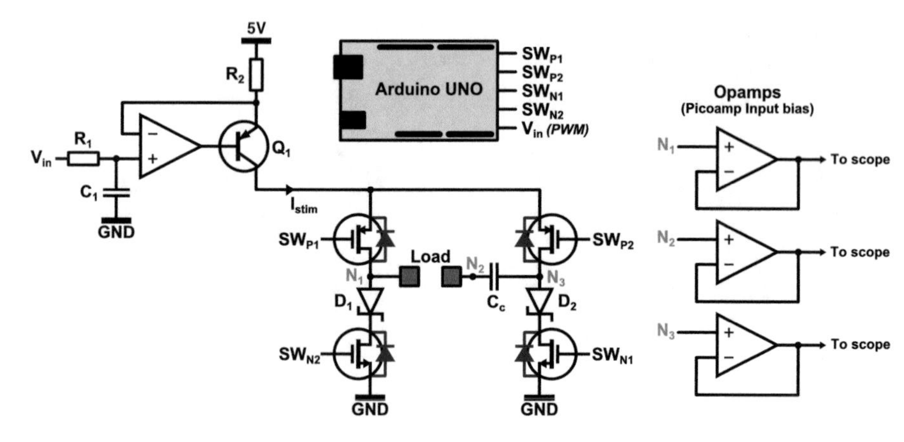

Fig. 3.4 Measurement setup used to verify the influence of the coupling capacitor C_c on the charge cancellation. A constant current source based around Q_1 is connected to the load via an H-bridge configuration (MOSFET switches), which allows bidirectional stimulation. An Arduino Uno is used for the control of the circuit, while buffers are used to prevent loading of the system during measurements

The value of C_c should be chosen well above C_{dl} in order to limit the contribution of C_c to the voltage headroom of the stimulator [21]. In this particular case it was chosen to make $C_c = 8.8\,\mu\text{F}$, based on the availability of components for the measurements. The circuit was simulated over many stimulation cycles (up to 200 s) to analyze the voltage over C_{dl} and C_c. To minimize leakage introduced by the simulator, the minimum conductance of the Spice simulator was lowered to $G_{min} =$ 1 fS. After a simulation, Matlab was used to select the time stamps that correspond to pseudo steady-state to obtain the values of V_{os} over many stimulation cycles.

After simulations, a stimulation circuit was built using discrete components as depicted in Fig. 3.4. Transistor Q_1 (2N3906) implements a current source together with resistor R_2 and the opamp (LMV358). The output current I_{stim} is controlled using the PWM signal V_{in} that is filtered using R_1 ($1\,\text{M}\Omega$) and C_1 ($1\,\mu\text{F}$). Using the H-bridge topology implemented with MOSFET devices (NTZD3155C), the current can be injected bidirectionally through the load during the cathodic and anodic stimulation phase. An Arduino Uno is used to control the switches: during the cathodic phase, switches SW_{P1} and SW_{N1} are closed, while during the anodic phase, switches SW_{P2} and SW_{N2} are closed. The tissue is shorted in between the stimulation pulses by closing SW_{P1} and SW_{P2}. Diodes D_1 and D_2 (CD0603-B00340) are needed to prevent unwanted current flow through the body diodes of SW_{N1} and SW_{N2} (indicated in dark-blue): if C_{dl} is charged beyond 0.6 V during the cathodic phase, the body diodes of SW_{N1} and SW_{N2} otherwise become forward biased when the stimulation direction is reversed.

The Arduino was programmed with four different stimulation settings as summarized in Table 3.1. The first setup uses no coupling capacitor and it is verified that C_{dl} is indeed charged back to 0 V after a stimulation cycle. The second setup uses

Table 3.1 Stimulation settings used during measurements

Nr	Waveform	I_{stim} (mA)	t_c (μs)	Mismatch η	f_{stim} (Hz)	Incl. C_c?
1	Biphasic	1.5	460	1.085 ($t_a = 500\,\mu$s)	110	No
2	Biphasic	1.5	460	1.085 ($t_a = 500\,\mu$s)	110	Yes
3	Biphasic	15	200	0.75 ($t_a = 150\,\mu$s)	400	Yes
4	Monophasic	15	200	0	100	Yes

a low intensity stimulation cycle with a positive charge mismatch, while the third setup uses a high intensity stimulation cycle (close to the maximum stimulation intensity possible before the current source would clip to the 5 V supply voltage). The load of the circuit in Fig. 3.4 first consisted of the electrode model from Fig. 4.7 ($R_s = 100\,\Omega$, $C_{dl} = 1.5\,\mu$F, and $R_{ct} = 1\,$MΩ). Subsequently the electrode model was replaced by the electrodes that were submerged in a PBS solution (pH 7.4, Gibco® Life Technologies™). The electrodes were stimulated in a bipolar fashion.

To measure the response of the system, the relevant output signals are buffered using picoampere input bias operational amplifiers (AD8625, powered with ±8 V) in order to prevent the measurement equipment from loading the system. It was found using simulations and measurements that a 10 MΩ||12 pF standard probe largely distorts the measurement, as will be discussed further in the last section.

3.2.2 Measurement Results

Figure 3.5 shows the simulation results of the circuit from Fig. 3.1. In Fig. 3.5a the value of V_{Cdl} in pseudo steady-state is shown over many stimulation cycles. When no coupling capacitor is used, V_{Cdl} can discharge almost completely. When C_c is added in Fig. 3.5a it is seen that after several stimulation cycles $V_{Cdl} = 20.7$ mV. Indeed the introduction of C_c causes an offset in V_{Cdl} in pseudo steady-state. Furthermore the simulated values correspond well with Eq. (3.4), which predicts $V_{os} = 20.6$ mV.

In Fig. 3.5a, b the simulated transient behavior of the voltages in the circuit including C_c is shown for two time instances. Figure 3.5b shows the voltages right after the first stimulation cycle, while Fig. 3.5c shows a stimulation cycle after 190 s of simulation time, where the offset is clearly visible.

In Fig. 3.6 the measurement results are presented for all experiments listed in Table 3.1. All waveforms were captured after stimulation was enabled sufficiently long (at least 5 min) to allow the voltages to settle. In all figures V_{out} refers to the voltage measured over the output of the current source (between nodes N_1 and N_3 in Fig. 3.4) and V_{el} is the voltage over the electrode (nodes N_1 and N_2). For saline measurements it is not possible to measure V_{Cdl} directly and hence V_{el} is shown instead.

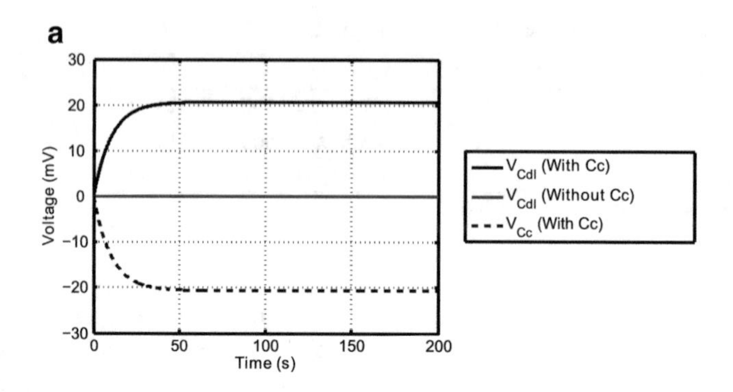

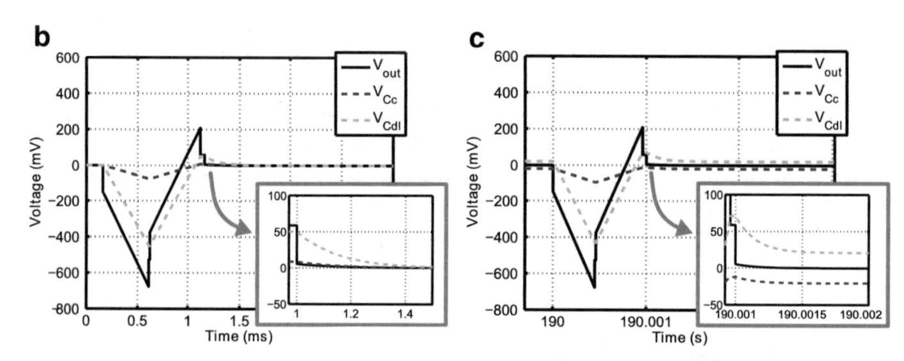

Fig. 3.5 Simulation results of the circuit from Fig. 3.1. In (**a**) the voltages V_{Cdl} and V_{Cc} are shown during the interval t_{open} over a large number of stimulation cycles. As can be seen, the coupling capacitor causes an offset that is possible due to the underdetermined Eq. (3.2). In (**b**) and (**c**) the transient voltages are shown for the system with C_c and $R_p = \infty$ just after stimulation is initiated and after 190 s, respectively

3.2.3 *Discussion*

The measured values for V_{os} are summarized in Table 3.2 and compared with the values calculated using Eq. (3.4). It is seen that the measurements with the model correspond well to the calculated values, indicating that the circuit implementation is working as expected. For the saline measurements the values of V_{os} are higher than expected and hence the model underestimates the offset value introduced. This is most likely due to complex non-linear behavior of the electrode–tissue interface that cannot be modeled using the simple capacitance C_{dl}. The electrode model is a small signal model (C_{dl} was found using a sinusoidal excitation of 0.1 V) and the measurement results show that the validity of the model is limited during a stimulation cycle. From these results and the plots in Fig. 6.12 we can draw three important conclusions.

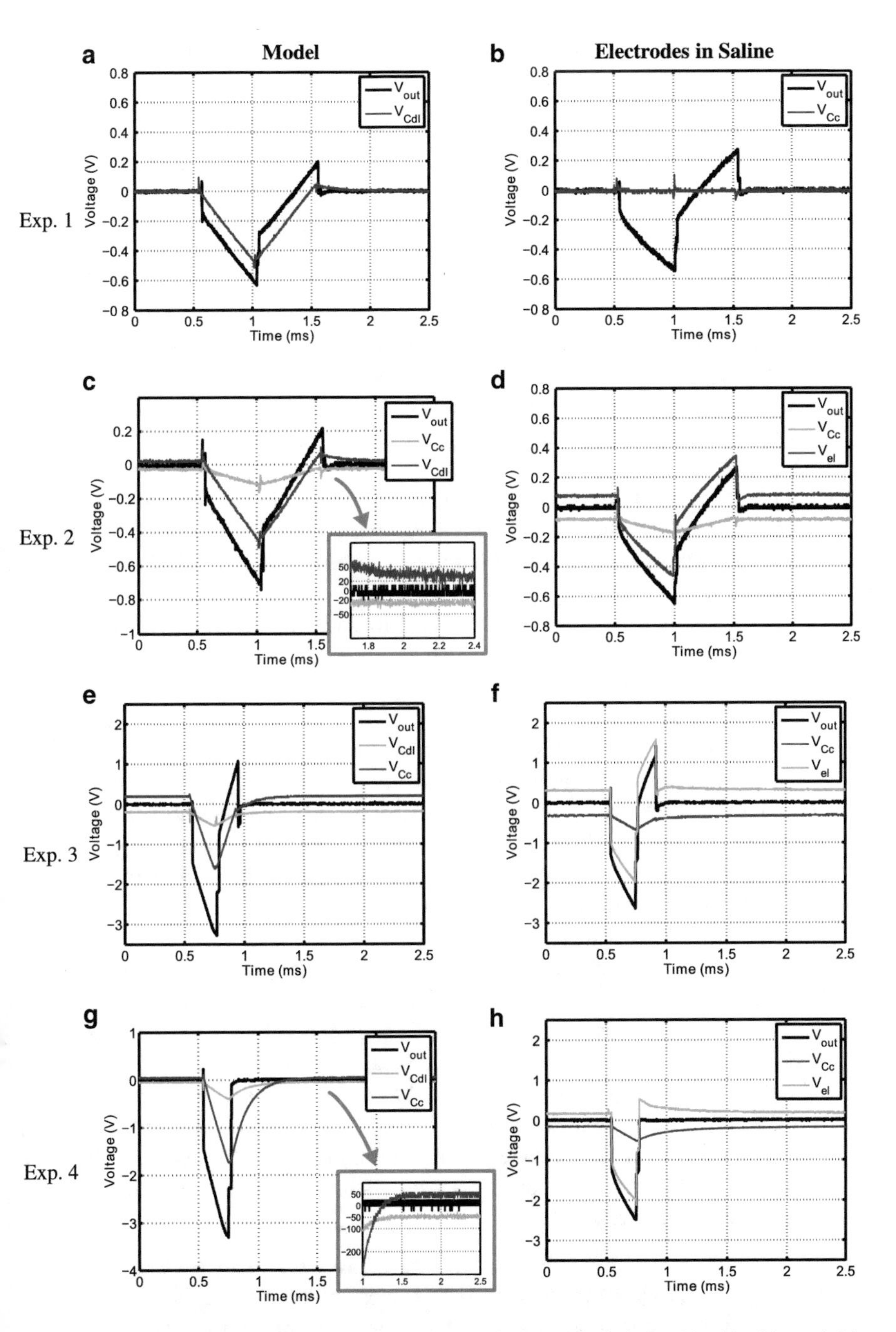

Fig. 3.6 Measurement results with the circuit described in Fig. 3.4. In (**a**), (**c**), (**e**), and (**g**) measurements with the electrode model are depicted according to the experiments listed in Table 3.1. (**b**), (**d**), (**f**), and (**h**) show the same measurements with electrodes in a saline bath. V_{out} is the voltage over nodes $N_1 - N_3$ (Fig. 3.4), while V_{el} is the voltage over nodes $N_2 - N_3$

Table 3.2 Calculated and measured values of V_{os} for the experiments summarized in Table 3.1

Experiment	Equation (3.4) (mV)	Measurement (model) (mV)	Measurement (saline) (mV)
1	0	0	0
2	21.6	25	80
3	201	200	320
4	50	50	165

(1) First of all, coupling capacitors barely improve the way in which V_{Cdl} returns to equilibrium. The only way in which C_c contributes is by making τ_{dis} (Eq. (3.1)) smaller during the t_{dis} interval [12]. This causes the interface to discharge towards equilibrium slightly faster. However, since $C_c \ll C_{dl}$, the influence on τ_{dis} is negligible and hence coupling capacitors barely improve the charge cancellation.

(2) Second of all, coupling capacitors introduce an offset in the pseudo-steady-state value of the electrodes. The value of V_{os} can be predicted using Eq. (3.4) and the validity has been confirmed with measurements with electrodes in a saline solution.

The question is whether or not V_{os} introduces potential safety issues. For small values of V_{os} no problems are expected: as long as no irreversible faradaic reactions are triggered, no harmful effects are to be expected. Even more so, V_{os} will increase the amount of charge that can be injected [22], because V_{os} reduces the peak voltage of V_{Cdl} during a stimulation cycle.

However, when V_{os} increases towards the threshold of irreversible faradaic reactions (600–900 mV for platinum electrodes [8]), problems can be expected. In this case the interface is experiencing a significant offset voltage during the t_{dis} interval, during which irreversible reactions might occur. For high stimulation intensities Fig. 3.3 predicts values of V_{os} that are close to or exceed the maximum safe voltage window.

(3) Finally, secondary effects can have a strong effect on V_{os}. In Fig. 3.7a, b measurements are presented with the settings of Experiment 2 and with the model and electrodes, respectively. This time the voltages were not buffered using the picoampere input bias opamps, but $10\,\text{M}\Omega || 12\,\text{pF}$ probes were connected to N_1, N_2, and N_3 directly. As can be seen this has a huge impact on the offset voltage: it increases to 2 and 0.6 V, respectively. These findings show that care has to be taken when additional circuitry such as electrode impedance monitors or recording amplifiers are added to a stimulator circuit that uses coupling capacitors.

All in all, it can be concluded that in contrast to what many other studies have suggested [12, 13, 21], the introduction of C_c does not improve the charge balancing process and it is furthermore associated with the loss of control over the pseudo steady-state value of V_{Cdl}. Instead of ensuring safety by returning the electrode interface voltage back to 0 V, the coupling capacitor introduces an unwanted offset in the interface voltage that is hard to control by the stimulator and, moreover, is sensitive to secondary effects.

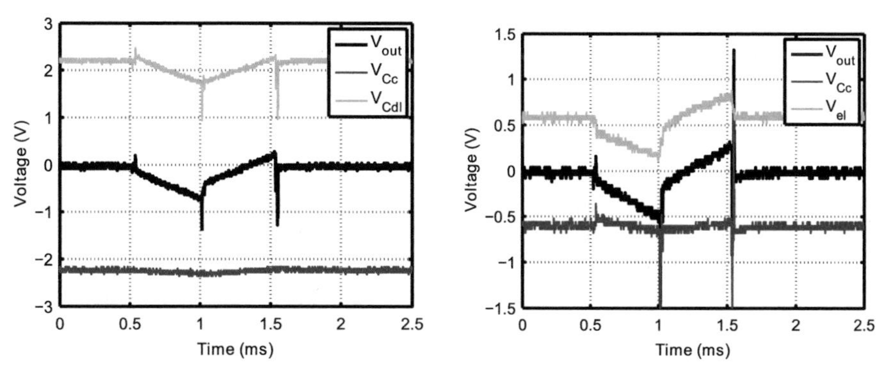

Fig. 3.7 Influence of probes in the measurement setup. In (**a**) the measurement results are shown with the electrode model connected for which the picoamp input bias opamps (Fig. 3.4) are removed. The value of V_{os} increases dramatically to more than 2 V. In (**b**) the same measurements are repeated with the electrodes in saline and this shows $V_{os} = 0.6\,\text{V}$

Although this work suggests that coupling capacitors are not beneficial for charge cancellation purposes, they still protect the electrodes and tissue from DC currents in case of a device or software failure. Depending on the application this could require the need to still use these capacitors. In that case the results from this study show that the stimulation settings should be limited to ensure that under all operating conditions V_{os} does not exceed any predefined safety window.

It is possible to discharge both C_c and C_{dl} completely by introducing an additional switch between the electrodes. In this case C_c and C_{dl} are shorted individually and are guaranteed to discharge towards 0 V in pseudo steady-state. However, in this case the coupling capacitor is not contributing in any way to the charge balancing process: it does not improve τ_{dis} and V_{Cdl} will have the same response as compared to the circuit without C_c. Furthermore the additional switch might introduce a single-fault device failure risk.

This work focused on passive charge balancing techniques. Active charge balancing techniques use feedback to bring the electrode voltage back to safe values after a stimulation cycle and can therefore help to overcome the offset problem. However, if these schemes require a coupling capacitor to protect in the event of a device failure, it is important to measure the voltage over the electrodes only and not to include the coupling capacitor. This requires an additional sensing pin if the coupling capacitors are realized using external components. Only then the feedback mechanism will help to remove the offset.

In this study, only one type of electrode was considered. Smaller electrodes have different impedance levels and more research is needed to find the pseudo steady-state response in this case. Note that Eq. (3.4) is only valid under the assumption that $\tau_{dis} \ll t_{dis}$, which might not be the case for high impedance electrodes. Finally, it would also be interesting to determine the influence of the coupling capacitors in vivo.

3.2.4 Conclusions

In this work the influence of coupling capacitors on the charge balancing properties is studied during neural stimulation. In contrast to what previous work suggests, coupling capacitors were found not to improve the charge balancing process. Even more so, they introduce an offset voltage in the electrodes, which cannot be removed by conventional means such as passive discharging. The value of the offset voltage depends on the stimulation and electrode parameters. When using coupling capacitors it is therefore important to ensure that this offset voltage does not exceed any safety boundaries for all possible operating conditions.

3.3 Reversibility of Charge Transfer Processes During Stimulation

As discussed in Sect. 3.1, electrochemically safe operation of polarizable electrodes during pulsatory electrical neurostimulation is important to prevent electrode and tissue damage. To prevent charge accumulation at the electrode–tissue interface of polarizable electrodes, traditionally charge balanced stimulation is applied. In its simplest form a biphasic stimulation pulse with equal charge contents is applied to the tissue. To ensure long term stability usually additional mechanisms are required, such as passive charge balancing, coupling capacitors (see previous section), or active charge balancing using feedback.

Many stimulator designs approach the problem of charge cancellation from the electrical domain and focus on designing perfectly matched current sources. It seems that most proposed methodologies assume that the electrodes are being operated within the charge injection limits for safe operation [7, 23]. These limits imply that all current through the electrode–tissue interface is transported using reversible charge transfer processes.

This section approaches the problem from the electrochemical domain. A methodology is proposed that aims to exclusively cancel the charge that has been transfered by means of reversible charge transfer processes. Any charge transfered by means of irreversible charge transfer processes should not be recovered. A feedforward mechanism is used to achieve this goal.

One of the advantages of the proposed method is that it can be used while the electrodes are being operated in their application. The method does not require additional electrodes (such as a standard (Ag-AgCl) reference electrode) and is active during the actual stimulation cycle. In this work bipolarly driven platinum electrodes with two typical sizes are submerged in a saline solution, but the principle is also suitable for in vivo setups and/or other types of electrode configurations.

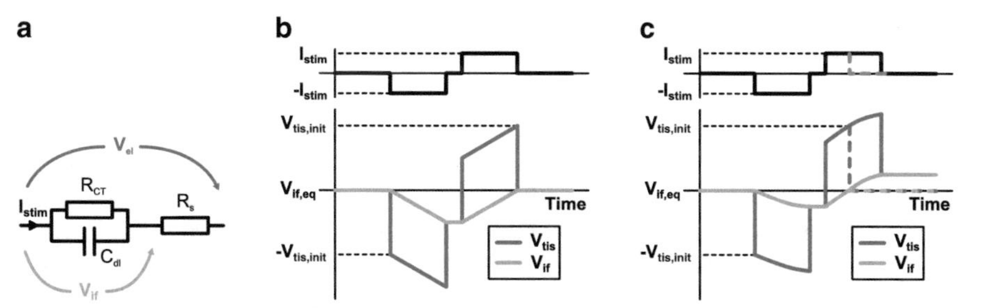

Fig. 3.8 In (**a**) the model for the electrodes in the saline solution that is used in this study is shown, along with the main electrical quantities. In (**b**) a typical response during a symmetrical biphasic stimulation waveform is shown for $R_{ct} \rightarrow \infty$. In (**c**) the same response is illustrated for when $R_{ct} \neq \infty$. Note that V_{if} returns to $V_{if,eq}$ when $V_{el} = V_{el,init}$, as shown by the *dashed line*

3.3.1 Theory

In Fig. 3.8a a model is shown that is used to analyze the response of the electrodes during a stimulation cycle [19]. The model consists of two parts: (1) the electrolyte (in this study a physiological saline solution) and (2) the interface between the electrode and the electrolyte. The electrolyte is assumed to be resistive and is modeled using the inter-cellular resistance ("spreading resistance") R_s [24].

The interface model has two parts. The reversible currents that pass through the electrode are modeled with C_{if}. Examples of such currents are purely capacitive currents due to charging of the dual-layer interface (common in noble metal electrodes), but also pseudo-capacitive currents such as surface-bound reversible faradaic processes (dominant in, e.g., AIROF electrodes). The second part is R_{ct}, which represents irreversible faradaic processes such as O_2 and H_2 evolution.

In Fig. 3.8b a schematic response of the electrodes using reversible charge transfer processes exclusively ($R_{ct} \rightarrow \infty$) during a symmetrical charge balanced biphasic stimulation cycle is depicted: the interface voltage V_{if} charges during the first cathodic stimulation phase and discharges back to the equilibrium voltage $V_{if,eq}$ during the second anodic stimulation phase.

In Fig. 3.8c it is shown what happens when the electrode current also includes an irreversible charge transfer component ($R_{ct} \neq \infty$): V_{if} is first of all charged at a lower rate during the cathodic phase due to the current through R_{ct}. Second of all V_{if} is discharged too much during the anodic phase, due to the fact that only the capacitive (reversible) current should be charge balanced, not the resistive (irreversible) current. Therefore $V_{if} \neq V_{if,eq}$ will now need to be reduced by charge balancing techniques, such as passive discharging (shortening of the electrode).

As can be seen from Fig. 3.8b, c, V_{if} returns back to $V_{if,eq}$ as soon as $V_{el} = V_{el,init}$, irrespective whether $R_{ct} \rightarrow \infty$ or not. Stopping the stimulation at this point (as indicated by the dashed lines in Fig. 3.8c) has two advantages. First of all, it will bring the interface back to $V_{if,eq}$, which eliminates the need for the interface to charge back during the electrode shortening [25]. Second of all, the symmetry in the duration of the cathodic and anodic phase (t_c and t_a, respectively) will now be a measure for the amount of irreversible processes during the stimulation cycle. When $R_{ct} \rightarrow \infty$, it will hold $t_c = t_a$, while in case of irreversible processes $t_c \neq t_a$.

This makes it possible to detect whether irreversible charge transfer occurs during a stimulation cycle, while the electrodes are being operated in the clinical application. There is no need for additional reference electrodes and the electrodes can be operated with the stimulation cycle that is required for the clinical application. The method relies on the fact that the voltage drop $I_{stim}R_s$ is constant during the stimulation cycle. Note that $V_{if,eq}$ does not necessarily need to be zero for the system to work. This means that the method can be used in combination with a DC electrode bias, which is used to increase the charge injection capacity of AIROF electrodes [26].

Also note that "flattening" of the slope (non-linear increase) in the response of Fig. 3.8c is not a sufficient condition to assume irreversible reactions. In a practical situation both C_{dl} and R_{ct} are likely to behave non-linear. This makes it possible for the slope to change without the need to trigger irreversible reactions. Note that the proposed method is independent of the slope and only relies on the $I_{stim}R_s$ voltage drop.

3.3.2 Methods

3.3.2.1 Stimulator Design

The principle illustrated in Fig. 3.8c is implemented in a discrete component circuit design as presented in Fig. 3.9. A constant current source is made with resistor R_2 over which the voltage is kept constant using the feedback loop around opamp OA_1. Transistor Q_1 conveys this current into the electrodes and the direction of current is changed using the H-bridge topology around MOSFETs M_1–M_4. High speed opamp OA_2 with an output enable (using the !SHDWN-pin) is used to sample the voltage $V_{el,init}$ at the beginning of a stimulation cycle on capacitor C_1, after which OA_2 is disabled again. This essentially implements a switched-opamp technique [27]. During the second stimulation phase opamp OA_3 will monitor the electrode voltage and comparator OA_4 compares this to the voltage stored on C_1. When the electrode voltage is equal to the capacitor voltage, this is detected by OA_4 and digital logic will stop the stimulation.

The circuit is implemented on a PCB. The supply voltage and control signals are 5 V and all come from an Arduino Uno microcontroller platform. The value of R_2 was chosen such that the voltage over this resistor was below 0.5 V, allowing enough voltage headroom for the load.

In [28] a system was proposed in which both the anodic and cathodic currents are limited based on the interface voltage of the working electrode. The idea is to limit the voltage swing over the interface and to return the interface back to equilibrium afterwards. However, this system needs an additional reference electrode, which

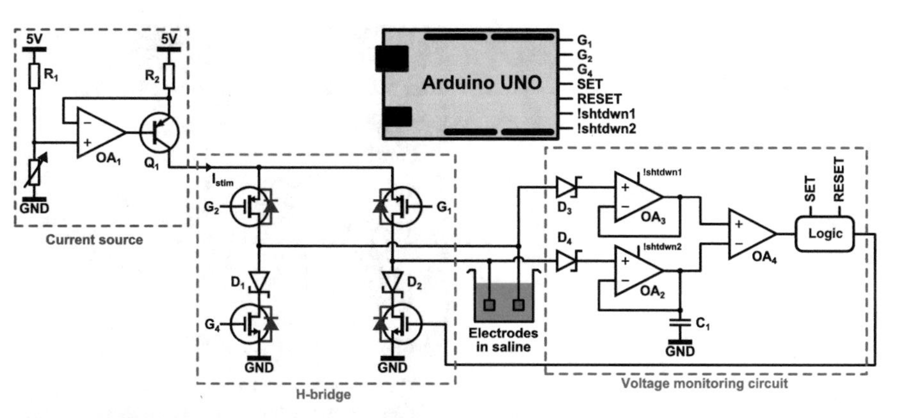

Fig. 3.9 Circuit implementation of the stimulation protocol. The current source generates a constant I_{stim} that is injected bidirectionally through the electrodes in the saline bath using the H-bridge M_1–M_4. The voltage monitoring circuit implements the feedforward control: the $I_{\text{stim}}R_{\text{spread}}$-drop is stored on C_1 during the cathodic phase using OA_2 that can be switched to a high impedance output using the !SHDWN-pin. During the anodic pulse OA_3 is used to compare the electrode voltage with the voltage stored on C_1. Logic is used to switch the stimulation off when comparator OA_4 detects the end of the stimulation cycle

makes the application in clinical settings more cumbersome. Furthermore, the current limiting circuitry creates a non-constant I_{stim}, which makes the circuit not suitable for our purpose.

3.3.2.2 Electrodes

Two different types of platinum electrodes are used for the measurements. The first type (see Fig. 3.10) is the same type of SCS electrodes ($R_s \approx 100\,\Omega$ and $C_{dl} \approx 1.5\,\mu\text{F}$) that were discussed in Sect. 3.2.1.The saline solution is a PBS solution (concentrations of 155.17 mM NaCl, 1.059 mM KH_2PO_4, and 2.97 mM Na_2HPO_4-$7H_2O$, pH 7.4, Gibco® Life technologies™).

The second type of electrodes used in this study were cochlear electrodes (Clarion® HiFocus™electrodes, Advanced Bionics) and are also depicted in Fig. 3.10. The electrode lead consist of 16 platinum contacts (approximately $300\,\mu\text{m}\times 500\,\mu\text{m} = 0.0015\,\text{cm}^2$) of which two were selected to drive the electrodes in a bipolar fashion. For these electrodes it was determined that $R_s \approx 1\,\text{k}\Omega$ and $C_{dl} \approx 40\,\text{nF}$ in the same PBS solution.

Stimulation intensities were first of all chosen such that the charge densities were kept below $50\,\mu\text{C/cm}^2$ to keep the operation below established charge injection limits for platinum electrodes [8]. Furthermore it was ensured that the current would not be limited by the supply voltage and the intensities were chosen to correspond with clinically used values for the type of electrodes [15, 26].

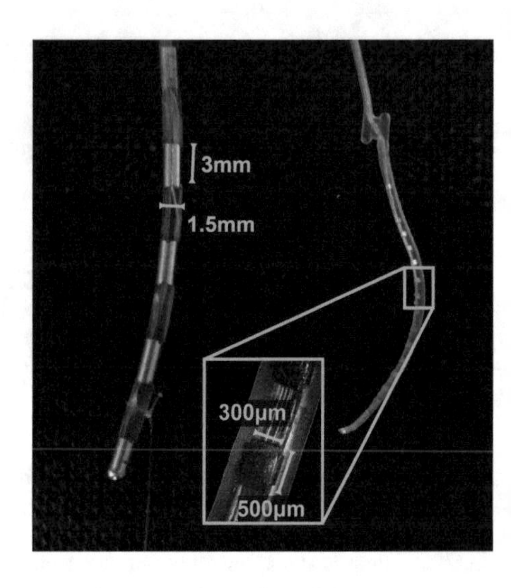

Fig. 3.10 The SCS electrodes (*left*) and the cochlear electrodes (*right*) that are used for the measurements. The SCS electrodes have an area of $0.14\,\text{cm}^2$, while the cochlear electrodes are $0.0015\,\text{cm}^2$

Both types of electrodes were stimulated with a pulsewidth $100\,\mu\text{s} < t_c < 1\,\text{ms}$ to investigate the response of both short and long stimulation pulses. For the SCS electrodes $I_{stim} < 1.5\,\text{mA}$, while for the cochlear electrodes $I_{stim} < 150\,\mu\text{A}$. For the highest stimulation current, the pulsewidth for the cochlear electrodes was limited to $400\,\mu\text{s}$ to limit the charge density.

The electrodes were stimulated with frequency $f_{stim} = 100\,\text{Hz}$ and were shorted in between the stimulation pulses. For each experiment the stimulation was enabled for 60 s before the first measurements were taken. After this initial settling time, the value of t_a was measured for 100 consecutive stimulation cycles and the median was used to compute the pulse duration ratio as t_a/t_c. The response of the electrodes was compared with a measurement in which the model of the electrodes (consisting of R_s and C_{dl}) is connected to the stimulator.

3.3.3 *Measurement Results*

Figure 3.11a shows an example of the measurement results for the SCS electrodes. The gray line depicts the response of the R_s-C_{dl} load, which is very similar to the response depicted in Fig. 3.8b. For this measurement $t_c = 400\,\mu\text{s}$, $I_{stim} = 1\,\text{mA}$. The black line depicts the response of the SCS electrodes, which is more similar to the response of Fig. 3.8c: the anodic phase reaches $V_{el,init}$ faster and therefore the ratio $t_a/t_c < 1$ for this particular case.

In Fig. 3.11b the pulse duration ratio t_a/t_c is plotted for a variety of I_{stim} and t_c for both the model and the electrodes. As expected the model gives an almost constant ratio close to 1, since all the stimulation current is used to charge and discharge C_{dl}. The electrodes show a dropping ratio for increasing I_{stim} and t_c.

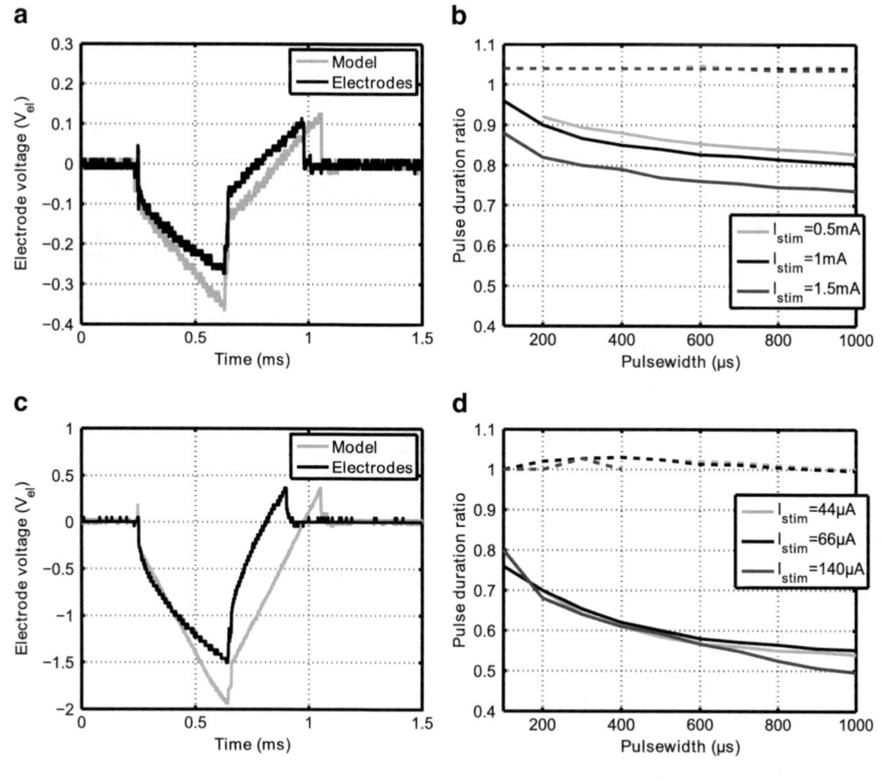

Fig. 3.11 Measurement results focusing on the t_a/t_c ratio for various stimulation settings and electrodes. In (**a**) an example of a measurement with the SCS electrodes (1 mA, 400 μs, 2.8 μC/cm^2) is shown for both the model as well as the electrodes, showing a clear difference in the t_a/t_c pulse duration ratio. In (**b**) an overview is given for the pulse duration ratio t_a/t_c for several values of I_{stim} and t_c. The *dashed lines* correspond to the measurements obtained from the model. In (**c**) and (**d**) the same plots are given for the cochlear electrodes

In Fig. 3.11c an example of a measurement result ($t_c = 400\,\mu s$, $I_{stim} = 140\,\mu A$) of the cochlear electrodes and their equivalent R_s-C_{dl} model is given. A similar result as with the SCS electrodes is obtained: the model has $t_a/t_c = 1$, while for the electrodes the ratio is lower. This is summarized in Fig. 3.11d in which again the model shows an almost constant response, while the ratio decreases for the electrodes.

In Fig. 3.12 all measurements from Fig. 3.11 are summarized as a function of the charge density. There is a clear trend that for increasing charge density the ratio t_a/t_c drops. Furthermore, this figure confirms that all measurements used a charge density below 50 μC/cm^2, corresponding to the reversible charge injection limit for platinum electrodes [8].

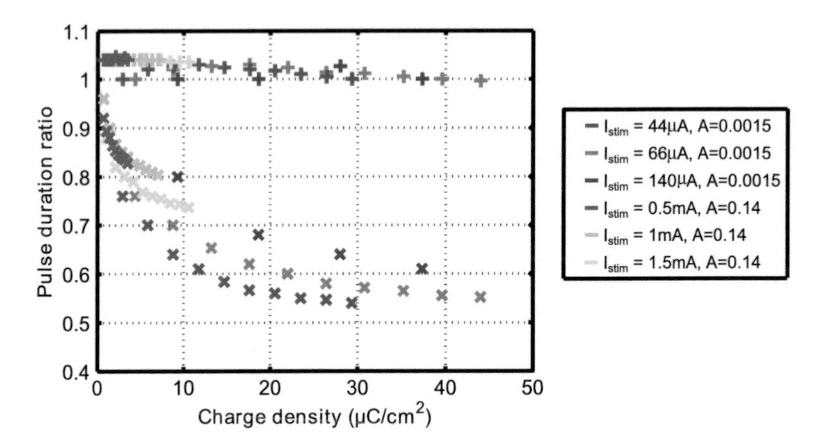

Fig. 3.12 Pulse duration ratio t_a/t_c as a function of the charge density for both the SCS electrodes and the cochlear electrodes. The + markers represent the ratio for the model, while the x markers represent the ratio for the electrodes in saline

3.3.4 Discussion

When using electrodes submerged in a saline solution, the ratio t_a/t_c drops significantly as is clear from Fig. 3.12. This can have various reasons. First of all it is possible that the circuit implementation is not working as expected. However, the measurement results with the model (ideal R_s-C_{dl} load) show that the proposed circuit is working as expected: in this case all the transported charge is reversible and the ratio t_a/t_c is therefore close to 1.

Another reason is that the model as used in Fig. 3.8a is not valid. The method heavily relies on the fact that the voltage drop over the tissue resistance $V_{tis} = I_{stim}R_s$ is constant. If this condition is not met, the stimulation would be stopped when the interface is not back at equilibrium.

To investigate this issue, the stimulation sequence is repeated, but without shortening the electrodes right after the stimulation cycle. It is now possible to see whether V_{if} returns to equilibrium after the stimulation cycle. In Fig. 3.13a the response of the SCS electrodes is shown for $I_{stim} = 1.5$ mA and $t_c = 400$ μs and it is compared to the case in which the feedforward control is disabled and ordinary charge balanced stimulation is used. Indeed the feedforward control causes V_{if} to return back to 0 V after the stimulation cycle, while the charge balanced stimulation does not. This indicates that for these electrodes and stimulation settings $I_{stim}R_s$ is constant.

Another explanation for the dropping ratio t_a/t_c is that there are irreversible charge transfer processes during the excitation of the electrodes. This would be a very surprising conclusion, since all stimulation settings were chosen well below the limits for reversible charge injection in platinum electrodes.

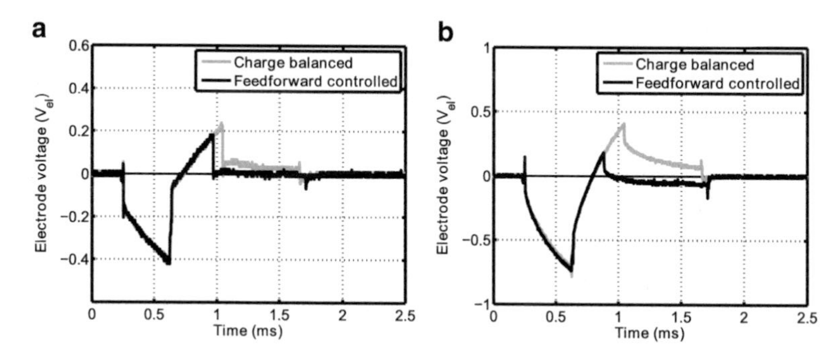

Fig. 3.13 Measurement results comparing the non-shorted electrode voltage right after the stimulation cycle in a charge balanced stimulation scheme and the proposed feedforward control scheme. In (**a**) the SCS electrode response is shown and the feedforward control returns the interface back to zero, in contrast to the charge balanced case. In (**b**) the response of the cochlear electrodes is shown and as can be seen the anodic phase of the feedforward control is slightly too short. In both plots the electrodes are shorted at approximately $t = 1.6\,\text{ms}$

One of the reasons that irreversible reactions might be triggered well below the reversible charge injection limit is that these limits are determined in an indirect fashion. First the interface voltage corresponding to the onset of irreversible reactions is determined using cyclic voltammetry [29]. Subsequently, a charge balanced biphasic stimulation pulse is imposed on the electrodes, while the interface voltage is monitored [30]. By preventing the interface voltage to exceed the voltage window that is considered as the border for irreversible reactions, the charge injection limit is determined. Due to the pulsatory excitation of the electrodes, the faradaic processes behave differently as compared to the pseudo static behavior during cyclic voltammetry. Due to the high rates of potential cycling the dynamic properties of the faradaic processes would need to be taken into account. As proposed in [8] it is likely that some faradaic processes do not have sufficient time to complete, which means that the charge transfer needs to be supported by other (possibly irreversible) processes.

Note that the question regarding the consequences for (long term) stimulation safety was not discussed. As shown in [1], electrochemical processes are not likely to be the most dominant factor in stimulation induced damage. Instead damage is mostly attributed to mechanical effects (leading to connective tissue formation and neuronal loss) or to stimulation induced effects (such as hyper-activation or electroporation [6]). Therefore, if indeed irreversible charge transfer mechanisms are triggered below the charge injection limits, it is not likely that these have significant impact to tissue or electrode damage.

3.4 Conclusions

This chapter investigated the electrode–tissue interface of polarizable electrodes during neural stimulation. The voltage over this interface needs to be well controlled to avoid potential problems due to electrochemical charge transfer processes.

First, the consequences for the interface voltage due to the use of a coupling capacitor were investigated. Despite many advantages in terms of safety, a coupling capacitor eliminates the control over the DC voltage over the electrode–tissue interface. It is shown that an offset voltage is created over this interface that could reach potentially dangerous levels.

Second, a new way of controlling the reversible charge during a stimulation cycle was introduced using a feedforward charge balancing technique. It was found that for platinum electrodes in a saline solution even for low stimulation intensities, charge imbalanced stimulation was needed for the electrode–tissue interface to return to equilibrium. The results suggest that irreversible charge transfer processes are triggered during a stimulation cycle.

References

1. Agnew, W.F., McCreery, D.B.: Considerations for safety with chronically implanted nerve electrodes. Epilepsia **31**(suppl.2), S27–S32 (1990)
2. Agnew, W.F., McCreery, D.B, Yuen, T.G.H., Bullara, L.A.: Histologic and physiologic evaluation of electrically stimulated peripheral nerve: considerations for the selection of parameters. Ann. Biomed. Eng. **17**(1), 39–60 (1989)
3. McCreery, D., Pikov, V., Troyk, P.R.: Neuronal loss due to prolonged controlled-current stimulation with chronically implanted microelectrodes in the car cerebral cortex. J. Neural Eng. **7**(3), 036005 (2010)
4. Agnew, W.F., McCreery, D.B., Yuen, T.G.H., Bullara, L.A.: Local anaesthetic block protects against electrically-induced damage in peripheral nerve. J. Biomed. Eng. **12**(4), 301–308 (1990)
5. Weaver, J.C., Chizmadzhez, Y.S.: Theory of electroporation: a review. Bioelectrochem. Bioenerg. **41**(2), 135–160 (1996)
6. Butterwick, A., Vankov, A., Huie, P., Freyvert, Y., Palanker, D.: Tissue damage by pulsed electrical stimulation. IEEE Trans. Biomed. Eng. **54**(12), 2261–2267 (2007)
7. Robblee, R.S., Rose, T.L.: Chapter 2. In: Agnew, W.F., McCreery, D.B. (eds.) Neural Prostheses: Fundamental Studies. Prentice Hall, New Jersey (1990)
8. Rose, T.L., Robblee, L.S.: Electrical stimulation with Pt electrodes. VIII. Electrochemically safe charge injection limits with 0.2 ms pulses. IEEE Trans. Biomed. Eng. **37**(11), 1118–1120 (1990)
9. Lilly, J.C., Hughes, J.R.: Brief, noninjurious electric waveform for stimulation of the brain. Science **121**, 468–469 (1955)
10. Prutchi, D., Norris, M.: Stimulation of excitable tissues. In: Design and Development of Medical Electronic Instrumentation: A Practical Perspective of the Design, Construction, and Test of Medical Devices. Wiley, New York
11. Parramon, J., Nimmagadda, K., Feldman, E., He, Y.: Multi-electrode implantable stimulator device with a single current path decoupling capacitor. US Patent 8,369,963 (2013)

12. Liu, X., Demosthenous, A., Donaldson, N.: Five valuable functions of blocking capacitors in stimulators. In: Proceedings of the 13th Annual International Conference of the FES Society (IFESS'08), pp. 322–324 (2008)
13. Sooksood, K., Stieglitz, T., Ortmanns, M.: An active approach for charge balancing in functional electrical stimulation. IEEE Trans. Biomed. Circuits Syst. **4**(3), 162–170 (2010)
14. Merrill, D.R., Bikson, M., Jefferys, J.G.R.: Electrical stimulation of excitable tissue – design of efficacious and safe protocols. J. Neurosci. Methods **141**, 171–198 (2005)
15. Site, J.J., Sarpeshkar, R.: A low-power blocking-capacitor-free charge-balanced electrode-stimulator chip with less than 6 nA DC error for 1-mA full-scale stimulation. IEEE Trans. Biomed. Circuits Syst. **1**(3), 172–183 (2007)
16. Nag, S., Jia, X., Thakor, N.V., Sharma, D.: Flexible charge balanced stimulator with 5.6 fC accuracy for 140 nC injections. IEEE Trans. Biomed. Circuits Syst. **7**(3), 266–275 (2013)
17. Liu, X., Demosthenous, A., Donaldson, N.: An integrated implantable stimulator that is fail-safe without off-chip blocking-capacitors. IEEE Trans. Biomed. Circuits Syst. **2**(3), 231–244 (2008)
18. van Dongen, M.N., Serdijn, W.A.: Does a coupling capacitor enhance the charge balance during neural stimulation? An empirical study. Med. Biol. Eng. Comput. 1–9 (2015). http://www.ncbi.nlm.nih.gov/pubmed/26018756
19. Malmivuo, J., Plonsey, R.: Bioelectromagnetism – Principles and Applications of Bioelectric and Biomagnetic Fields. Oxford University Press, New York (1995)
20. Eon Rechargeable IPG tech specs. St. Jude Medical Inc. http://professional.sjm.com/products/neuro/scs/generators/eon-rechargeable-ipg#tech-specs (2013). Cited 14 July 2014
21. Sooksood, K., Stieglitz, T., Ortmanns, M.: An experimental study on passive charge balancing. Adv. Radio Sci. **7**, 197–200 (2009)
22. Donaldson, N.N., Donaldson, P.E.K.: When are actively balanced biphasic (Lilly) stimulating pulses necessary in a neurological prosthesis? II pH changes; noxious products; electrode corrosion; discussion. Med. Biol. Eng. Comput. **24**(1), 50–56 (1986)
23. Cogan, S.F.: Neural stimulation and recording electrodes. Ann. Rev. Biomed. Eng. **10**, 275–309 (2008)
24. Wiertz, R.W.F., Rutten, W.L.C., Marani, E.: Impedance sensing for monitoring neuronal coverage and comparison with microscopy. IEEE Trans. Biomed. Eng. **57**(10), 2379–2385 (2010)
25. Woods, V.M., Triantis, I.F., Toumazou, C.: Offset prediction for charge-balanced stimulus waveforms. J. Neural Eng. **8**(4), 046032 (2011)
26. Cogan, S.F., Troyk, P.R., Ehrlich, J., Plante, T.D.: In vitro comparison of the charge-injection limits of activated iridium oxide (AIROF) and platinum-iridium microelectrodes. IEEE Trans. Biomed. Eng. **52**(9), 1612–1614 (2005)
27. Crols, J., Steyaert, M.: Switched-Opamp: an approach to realize full CMOS switched-capacitor circuits at very low power supply voltages. IEEE J. Solid-State Circuits **29**(8), 936–942 (1994)
28. Troyk, P.R., Detlefsen, D.E., Cogan, S.F., Ehrlich, J., Bak, M., McCreery, D.B., Bullara, L., Schmidt, E.: Safe charge-injection waveforms for iridium oxide (AIROF) microelectrodes. In: Proceedings of the 26th Annual International Conference of the IEEE Engineering in Medicine and Biology Society, pp. 4141–4144 (2004)
29. Brummer, S.B., Turner, M.J.: Electrochemical considerations for safe electrical stimulation of the nervous system with platinum electrodes. IEEE Trans. Biomed. Eng. **24**(1), 59–63 (1977)
30. Cogan, S.F., Troyk, P.R., Ehrlich, J., Gasbarro, C.M., Plante, T.D.: The influence of electrolyte composition on the in vitro charge-injection limits of activated iridium oxide (AIROF) stimulation electrodes. J. Neural Eng. **4**(2), 79–86 (2007)

Chapter 4
Efficacy of High Frequency Switched-Mode Neural Stimulation

Abstract This chapter investigates the efficacy of a fundamentally different neural stimulation technique: high frequency switched-mode neural stimulation. Instead of using a constant stimulation amplitude, the stimulus is switched on and off repeatedly with a high frequency (up to 100 kHz) duty cycled signal. It is first shown that switched-mode stimulation depolarizes the cell membrane in a similar way as classical constant amplitude stimulation.

As a first step tissue modeling that includes the dynamic properties of both the tissue material as well as the axon membrane is used to compare the activation mechanism of switched-mode stimulation with classical constant amplitude stimulation. These findings are subsequently verified using in vitro experiments in which the response of a Purkinje cell is measured due to a stimulation signal in the molecular layer of the cerebellum of a mouse. For this purpose a stimulator circuit is developed that is able to produce a monophasic high frequency switched-mode stimulation signal.

4.1 Introduction

Traditional functional electrical stimulation typically uses a current source with constant amplitude I_{stim} and pulsewidth t_{pulse} to recruit neurons in the target area. Early stimulator designs consisted of relatively simple programmable current source implementations. Over the years numerous modifications have been proposed to improve important aspects such as power efficiency, safety, and size. Most stimulators, however, still use constant current at the output.

Several implementations have investigated the use of alternative stimulation waveforms in an attempt to improve the performance. Some implementations focus on improving the efficiency of the activation mechanism in the neural tissue [1, 2]. Others focus on increasing the performance of the stimulator itself, of which several studies have proposed the use of high frequency stimulation waveforms. In [3] a 250 kHz pulsed waveform is used to decrease the size of the coupling capacitors. Two of these waveforms are subsequently added in antiphase to reconstruct a conventional stimulation waveform. In [4] a 10 MHz forward-buck and reverse-

M. van Dongen, W. Serdijn, *Design of Efficient and Safe Neural Stimulators*, Analog Circuits and Signal Processing, DOI 10.1007/978-3-319-28131-5_4

boost converter is used to increase the power efficiency of the stimulator by using inductive energy recycling. External capacitors are used to low-pass filter the switched signal and reconstruct a conventional waveform.

In Chap. 7 the design of a neural stimulator frontend is presented that uses a high frequency signal to stimulate the tissue. This stimulator brings several advantages as compared to constant current stimulation, such as high power efficiency, low number of external components, and support for complex electrode configurations. In this chapter the electrophysiological feasibility of a high frequency stimulator is investigated both theoretically as well as experimentally by determining whether a high frequency stimulation signal can indeed induce neural recruitment in a similar fashion as during classical constant current stimulation [5].

The high frequency stimulation pattern that is used to stimulate the tissue is assumed to be square shaped. The schematic circuit diagrams of both voltage and current based stimulation are depicted in Fig. 4.1a. A fixed value for V_{stim} or I_{stim} is used, while the stimulation intensity is controlled by driving the switch with a pulse width modulated (PWM) signal; this is referred to as switched-mode operation. In Fig. 4.1b a sketch is given of the monophasic stimulation pulse resulting from either of the circuits. The switch is operated with duty cycle δ and switching period t_s. This results in an average stimulation intensity $V_{avg} = \delta V_{stim}$ or $I_{avg} = \delta I_{stim}$ for voltage and current based stimulation, respectively.

It is important to note that in this work the term ‘high frequency’ refers to the frequency of the pulses that make up a single stimulation waveform. It does not refer to the repetition rate at which the stimulation cycles are repeated. Furthermore, this work investigates the electrophysiological feasibility of switched-mode stimulator circuits; it does not aim to design a stimulation waveform that improves the activation mechanism itself with respect to classical constant current stimulation.

The organization of this chapter is as follows. In Sect. 4.2 the tissue and the cell membrane are modeled with frequency dependent parameters. These models are used to analyze the response of the membrane voltage to the high frequency stimulation signal. In Sect. 4.3 the experimental setup is discussed, consisting of a prototype high frequency stimulator in combination with an in vitro patch clamp recording setup. Finally in Sects. 4.4 and 4.5 the measurement results are presented and discussed.

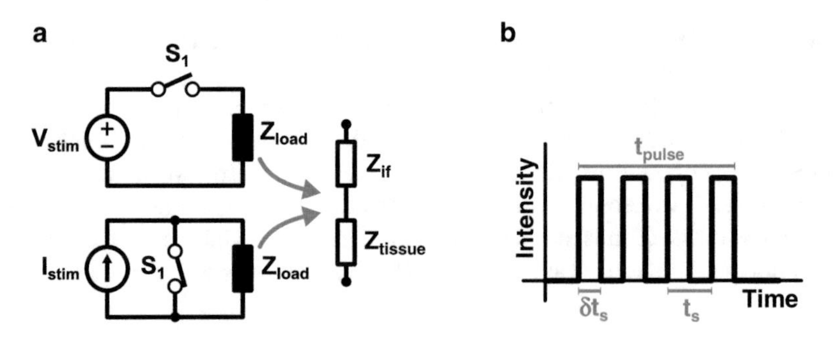

Fig. 4.1 (**a**) Schematic representation of a high frequency voltage and current system that is driven by a switch that is controlled by a PWM signal. In (**b**) the resulting stimulation signal is sketched

4.2 Theory

A high frequency (switched) signal that is injected via the electrodes will be filtered by the tissue. First the tissue material properties influence the transient voltage over and current through the tissue. Subsequently the electric field in the tissue and the properties of the cell membrane will determine the transient shape of the membrane voltage, which is ultimately responsible for the actual activation or inhibition of the neurons. These two processes will be discussed separately.

4.2.1 Tissue Material Properties

In Fig. 4.1a the tissue is modeled with Z_{if} (interface impedance) and Z_{tis} (tissue impedance). For current based stimulation $V_{tis} = I_{stim}Z_{tis}$ is independent of Z_{if}. For voltage based stimulation $V_{tis} = V_{stim} - V_{if}$ with V_{if} the voltage over Z_{if}. In this study non-polarizable Ag/AgCl electrodes will be used for which $Z_{if} \approx 0$ and therefore $V_{tis} \approx V_{stim}$ [6].

The tissue voltage V_{tis} and current I_{tis} are related to each other via the resistive and reactive properties of the tissue. In [7] the capacitive and resistive properties of the tissue are measured for a wide range of frequencies and human tissue types. The resistivity and permittivity of gray matter as a function of the frequency are plotted in Fig. 4.2a. This plot has been obtained by calculating the relative permittivity ϵ_r and conductivity σ based on the equation for the relative complex permittivity $\hat{\epsilon}_r(\omega)$ from [7]:

$$\epsilon_r(\omega) = \text{Re}\left[\hat{\epsilon}_r(j\omega)\right] \tag{4.1}$$

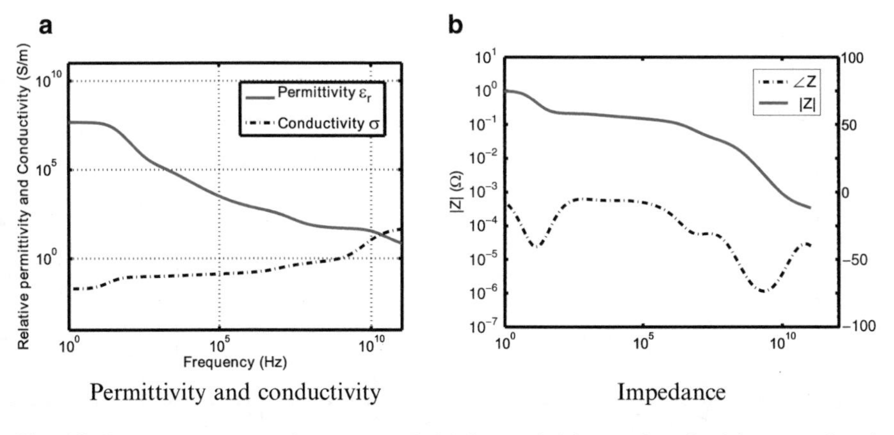

Fig. 4.2 Frequency response of gray matter. In (**a**) the permittivity ϵ and conductivity σ are plotted as function of the frequency [7] and in (**b**) the corresponding normalized (impedance) Bode plots are given

$$\sigma(\omega) = \text{Im}\left[\hat{\epsilon}_r(j\omega)\right] \cdot -\epsilon_0\omega \tag{4.2}$$

Here ϵ_0 is the permittivity of free space. As can be seen neural tissue shows strong dispersion for $\hat{\epsilon}_r(\omega)$. To find the relation between the tissue voltage and current the values of ϵ_r and σ need to be converted to impedance. Given $\hat{\epsilon}_r$ the impedance Z is:

$$Z = \frac{1}{\hat{\epsilon}_r j\omega C_0} \tag{4.3}$$

Here C_0 is a constant that sets the absolute value of the impedance, which depends among other things on the electrode geometry. The impedance can be normalized, such that $|Z(0)| = 1$ by using:

$$\lim_{\omega\to 0} |Z(j\omega)| = \lim_{\omega\to 0} \left[\sqrt{(\epsilon_r(\omega))^2 + \left(\frac{-\sigma(\omega)}{\omega\epsilon_0}\right)^2}\,\omega C_o\right]^{-1} = \frac{\epsilon_0}{\sigma(0)C_0} \tag{4.4}$$

Here Eqs. (4.1) and (4.2) are substituted for $\hat{\epsilon}_r(\omega)$ and $\sigma(0)$ is the conductance of the tissue at $\omega = 0$. From this it follows that $C_0 = \epsilon_0/\sigma(0)$ to normalize the transfer such that $|Z(0)| = 1$. The Bode plots of this normalized impedance are given in Fig. 4.2b.

This plot can now be used to obtain the shape for I_{tis} and V_{tis} and, if the impedance of the tissue is known for a certain frequency, it can be scaled to obtain the correct absolute values.

As an example, a 100 µA, 200 kHz, $\delta = 0.4$ switched-current signal $i_{in}(t)$ is supplied to an electrode system that has an impedance of $|Z| = 10\,\text{k}\Omega$ at 1 kHz. The tissue voltage is now found by solving $V_{out}(t) = \mathscr{F}^{-1}\left[Z \cdot \mathscr{F}\left[i_{in}(t)\right]\right]$, which is plotted in Fig. 4.3a. Indeed the tissue voltage is filtered and in the next section it will be seen that this is important for determining the activation of the neurons.

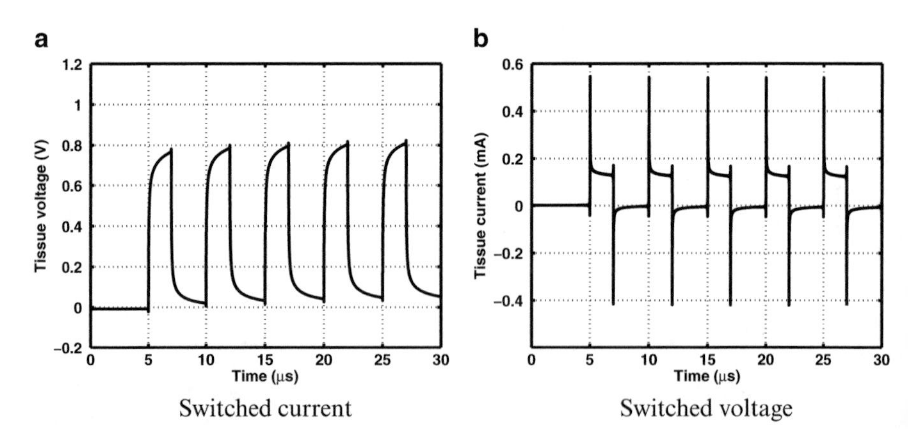

Fig. 4.3 The response V_{tis} to a square wave current input (**a**) and the response I_{tis} to a square wave voltage input (**b**), based on the impedance as given in Fig. 4.2b

Similarly a 1 V, 200 kHz, $\delta = 0.4$ switched-voltage signal $v_{in}(t)$ can be applied. The tissue current follows from $I_{out}(t) = \mathscr{F}^{-1}\left[\mathscr{F}\left[v_{in}(t)\right]/Z\right]$ and is plotted in Fig. 4.3b. The current spikes in Fig. 4.3b are due to the rapid charging of the capacitive properties of the tissue that arise from $\epsilon_r(\omega)$.

4.2.2 *Tissue Membrane Properties*

After the transient intensities of the tissue voltage and current have been determined by the stimulation protocol and the tissue impedance, it can be investigated how these quantities influence the neurons. Analogous to [8] the activation of the neurons is considered in the axons, for which the membrane voltage is determined using the cable equations. For these equations first the potential in the tissue as a function of the distance from the electrode is needed. When the electrode is considered to behave as a point source at the origin, the tissue potential has a $1/r$ dependence assuming quasi-static conditions [8]: $\Phi(r) = I_{stim}/(\sigma 4\pi r)$, where r is the distance from the electrode.

In [9] the influence on the tissue potential due to high frequency components in the stimulation signal was analyzed. It was found that the propagation effect was negligible and that only the complex permittivity as discussed in the previous section was significant. To incorporate these properties, the potential $\Phi(r, j\omega)$ can be determined in the frequency domain by substituting σ with the complex permittivity, leading to:

$$\Phi(r, j\omega) = \frac{I_{tis}(j\omega)}{j\omega\epsilon_0\hat{\epsilon}_r 4\pi r} \tag{4.5}$$

By transforming this potential back to the time domain, the transient of the potential $\Phi(r,t) = \mathscr{F}^{-1}\left[\Phi(r, j\omega)\right]$ at any distance r from the point source is obtained. Since the current is divided by the complex permittivity and since there are no propagation effects, the transient shape of the potential is proportional to V_{tis} as obtained in Fig. 4.3: it is just scaled as a function of the distance.

Next $\Phi(r,t)$ can be used as an input for an axon model to determine the response of the membrane voltage. The electrical parameters that are used for the axon model are summarized in Table 4.1 [8, 10]. In the following section the response of both myelinated and unmyelinated axons is considered. In both sections the fiber diameter is chosen to be small ($d_o = 0.8\,\mu\text{m}$), based on the axon diameter of unmyelinated Purkinje cells, which will be used later in the in vitro experiments [11].

4.2.2.1 Myelinated Axons

For a myelinated axon the model in Fig. 4.4a is used. The myelinated parts of the axon do not have ionic channels and are therefore modeled using the intracellular

Table 4.1 Axon properties used for the axon model [8, 10]

Symbol	Description	Value
ρ_i	Axoplasm resistivity	$54.7\ \Omega\ \text{cm}$
ρ_o	Extracellular resistivity	$0.3\ \text{k}\Omega\ \text{cm}$
c_m	Nodal membrane capacitance/unit area	$2.5\ \mu\text{F/cm}^2$
ν	Nodal gap width	$1.5\ \mu\text{m}$
l/d_o	Ratio of internode spacing to fiber diameter	100
d_i/d_o	Ratio of axon diameter to fiber diameter	0.6
g_{Na}	Sodium conductance/unit area	$120\ \text{mS/cm}^2$
V_{Na}	Sodium reversal voltage	115 mV
g_K	Potassium conductance/unit area	$36\ \text{mS/cm}^2$
V_K	Potassium reversal voltage	−12 mV
g_L	Leakage conductance/unit area	$0.3\ \text{mS/cm}^2$
V_L	Leakage voltage	10.61 mV

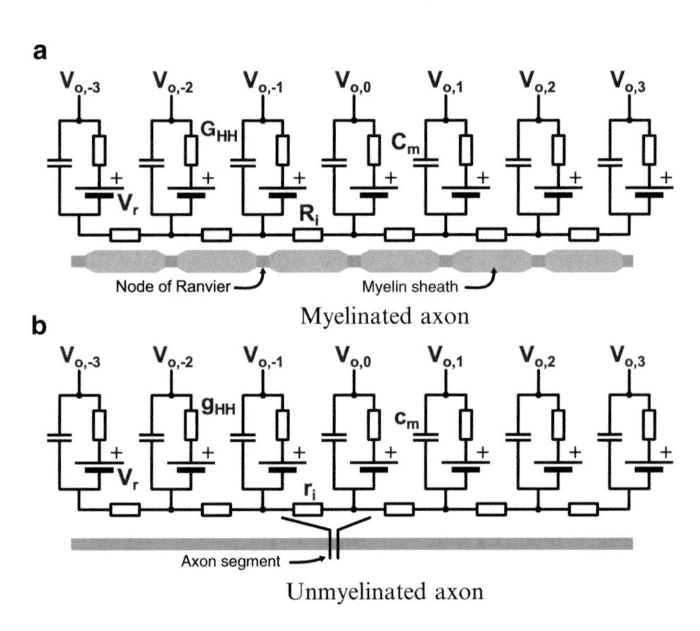

Fig. 4.4 Axon models for a myelinated axon (**a**) and an unmyelinated axon (**b**), used to find the response of the membrane voltage. The tissue potential at nodes V_1-V_4 is substituted after which the membrane voltage is found using circuit simulations

resistance $R_i = \rho_i l/(\pi(d_i/2)^2$, in which l represents the internode spacing and d_i the axon diameter. At the nodes of Ranvier the membrane is characterized by the membrane capacitance $C_m = c_m \pi d_i \nu$, the rest potential $V_{rest} = -70\,mV$, and the nonlinear conductance G_{HH}. The current through this conductance is given by the Hodgkin Huxley equations [12].

The membrane voltage $V_{m,n}$ at node n can be found by solving the following equation that follows directly from Kirchhoff's laws [8]:

$$\frac{dV_{m,n}}{dt} = \frac{1}{C_m}\left[\frac{1}{R_i}(V_{m,n-1} - 2V_{m,n} + V_{m,n+1} + V_{o,n-1} - 2V_{o,n} + V_{o,n+1}) - \pi d_i v i_{HH}\right] \tag{4.6}$$

Here $V_{o,n}$ is the voltage due to the electric field at node n that follows from Eq. (4.5) and i_{HH} is the current density given by the Hodgkin Huxley equations:

$$i_{HH} = g_{Na}m^3h(V_{m,n} - V_{rest} - V_{Na}) + g_K n^4(V_{m,n} - V_{rest} - V_K) + g_L(V_{m,n} - V_{rest} - V_L) \tag{4.7}$$

$$\frac{dm}{dt} = \alpha_m(1-m) - \beta_m m \tag{4.8}$$

$$\frac{dh}{dt} = \alpha_h(1-h) - \beta_h h \tag{4.9}$$

$$\frac{dn}{dt} = \alpha_n(1-n) - \beta_n n \tag{4.10}$$

The conductances g_{Na}, g_K, and g_L as well as the voltages V_{Na}, V_K, and V_L are constants, while α_x and β_x depend on the membrane voltage $V' = V_m - V_{rest}$ via:

$$\alpha_m = \frac{0.1 \cdot (25 - V')}{\exp\frac{25-V'}{10} - 1} \quad \alpha_h = \frac{0.07}{\exp\frac{V'}{20}} \quad \alpha_n = \frac{0.01(10 - V')}{\exp\frac{10-V'}{10} - 1} \tag{4.11a}$$

$$\beta_m = \frac{4}{\exp\frac{V'}{18}} \quad \beta_h = \frac{1}{\exp\frac{30-V'}{10} + 1} \quad \beta_n = \frac{0.125}{\exp\frac{V'}{80}} \tag{4.11b}$$

The response of the membrane potential due to the high frequency electric field can now be found by solving the differential equations above. This is done in Matlab by using the classical Runge–Kutta method (RK4). A step size of 1 μs is chosen during the high frequency stimulation interval, while after the stimulation pulse a step size of 10 μs is used.

A switched-voltage stimulation scheme with $V_{stim} = 1\,\text{V}$, $|Z(1\,\text{kHz})| = 1\,\text{k}\Omega$, $\delta = 0.5$, $f_{stim} = 1/t_s = 100\,\text{kHz}$, and $t_{pulse} = 100\,\mu\text{s}$ was chosen first. An axon with the center node at a distance $y = 0.5\,\text{mm}$ was considered. For this axon $C_m = 56.6\,\text{fF}$, $R_i = 241.8\,\text{M}\Omega$, the nodes of Ranvier are spaced 80 μm apart and a total of 9 nodes were simulated.

The resulting membrane voltage is depicted in Fig. 4.5a. First, the effect of the switched-mode stimulation can clearly be seen in the staircase transient shape of the membrane voltage. Furthermore it can be seen that the increase in the membrane voltage also leads to an action potential in the axon. This shows that according to the models, switched-mode stimulation can induce activation in the axons. Finally, this action potential is able to travel along the axon, as is shown by

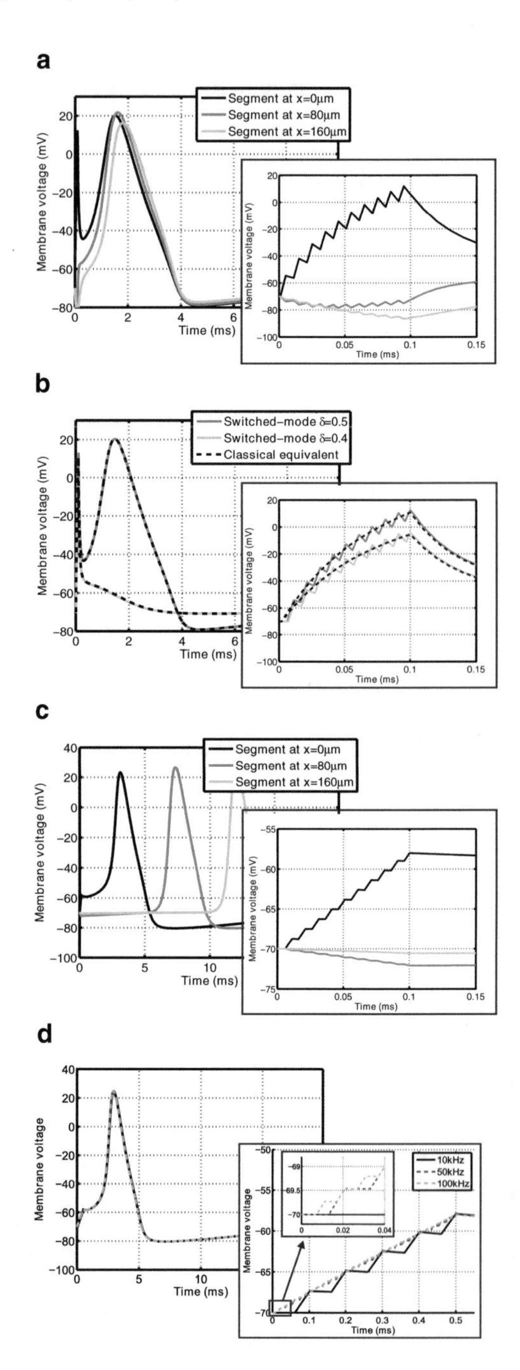

Fig. 4.5 Transient membrane voltage due to switched-mode stimulation according to the models of Fig. 4.4 for a variety of settings. In (**a**) the membrane voltage at three nodes of Ranvier of a myelinated axon is depicted during and after stimulation with a $\delta = 0.5$ switched-voltage source for which an action potential is generated. In (**b**) the effect of intensity (duty cycle for switched-mode versus amplitude for classical stimulation) is depicted and compared. In (**c**) the response at three points in an unmyelinated axon is shown, where it is shown that it is also possible to create action potentials. In (**d**) the response of an unmyelinated axon is given for $f_{stim} = 10,\ 50,$ and 100 kHz ($\delta = 0.4$), which shows that f_{stim} has no significant influence on the activation. In all plots a zoom is given of the membrane voltage during the stimulation

the response of the other nodes of Ranvier in Fig. 4.5a. A very similar result can be obtained when using switched-current stimulation.

In Fig. 4.5b the effect of the duty cycle δ is shown. The dark line shows the response for $\delta = 0.5$ and the light line is the response for $\delta = 0.4$. The latter setting is not able to induce an action potential anymore, which shows that δ is an effective way of controlling the stimulation intensity. The response is compared with a classical constant voltage stimulation with $V_{stim,classical} = \delta V_{stim}$ and is indicated with the dashed lines. Indeed an equivalent response is found.

4.2.2.2 Unmyelinated Axons

For unmyelinated axons the model as depicted in Fig. 4.4b is used. The axon is now divided into segments of length Δx with each segment containing an intracellular resistance per unit length: $r_i = 4\rho_i/d_i$, the capacitance per unit area c_m, the resting potential $V_{rest} = -70\,\text{mV}$, and the ionic conductance per unit area g_{HH}. Again a differential equation can be found that solves the membrane voltage $V_{m,n}$ [13]:

$$\frac{dV_{m,n}}{dt} = \frac{1}{c_m}\left[\frac{(V_{m,n-1} - 2V_{m,n} + V_{m,n+1}}{r_i(\Delta x)^2} + \frac{V_{o,n-1} - 2V_{o,n} + V_{o,n+1}}{r_i(\Delta x)^2} - i_{HH}\right] \tag{4.12}$$

An unmyelinated axon is considered at a distance $y = 0.5\,\text{mm}$. The axon is divided into 501 segments of 1 μm and has an outer diameter $d_o = 0.8\,\mu\text{m}$. For unmyelinated axons a higher stimulation intensity is needed in order to get effective stimulation. A voltage-mode stimulation signal with $V_{stim} = 10\,\text{V}$ and $\delta = 0.5$ is used.

The same solving strategy is chosen to solve Eq. (4.12). The membrane potential is depicted in Fig. 4.5c and looks very similar to the myelinated response. Also in this case the action potential is able to travel along the axon as shown by the response of segments that are further down the axon. Note that the propagation speed is much lower than in the myelinated case, which is a well known property.

Figure 4.5d shows the effect of f_{stim}: frequencies of 10, 50, and 100 kHz are used. As can be seen both the membrane voltage after the stimulation pulse and the response of the tissue do not depend on f_{stim}.

The simulation results show that switched-mode stimulation is able to induce the same sort of activation as classical stimulation in both myelinated as well as unmyelinated axons. The duty cycle δ is used to control the stimulation intensity in exactly the same way as the amplitude for classical stimulation. Note that compared to the tissue material properties the membrane time constant is much larger and is therefore dominant in the filtering process.

4.3 Methods

In order to verify whether the proposed high frequency stimulation scheme is able to induce neuronal recruitment by using the tissue filtering properties, an in vitro experiment is performed.

4.3.1 Recording Protocol

The in vitro recordings were taken in brain slices from the vermal cerebellum of C57Bl/6 inbred mice using a method similar to [14]. In short, mice were decapitated under isoflurane anesthesia and subsequently the cerebellum was removed and parasagittally sliced to preserve the Purkinje cell dendritic trees (250 μm thickness) using a Leica vibratome (VT1000S). The slice was kept for at least 1 h in artificial cerebrospinal fluid (ACSF) containing the following (in mM): 124 NaCl, 5 KCl, 1.25 Na_2HPO_4, 2$MgSO_4$, 2$CaCl_2$, 26 NaHCO, and 20D-glucose, bubbled with 95 % O_2 and 5 % CO_2 at 34 °C. 0.1 mM picrotoxin was added to the ACSF to block the inhibitory synaptic transmission from the molecular layer interneurons. This allows for recording of postsynaptic responses in the Purkinje cells due to stimulation of the granular cell axons.

Experiments were carried out under a constant flow of oxygenated ACSF at a rate of approximately 2.0 ml/min at 32±1 °C. The Purkinje cells were visualized using an upright microscope (Axioskop 2 FS plus; Carl Zeiss) equipped with a 40x water-immersion objective.

The stimulus electrode is an Ag-AgCl electrode in a patch pipette pulled from borosilicate glass (outer diameter 1.65 mm and inner diameter 1.1 mm) and is filled with ACSF. This electrode has an impedance $Z_{tis} \approx 3\,\mathrm{M}\Omega$ and is stimulated

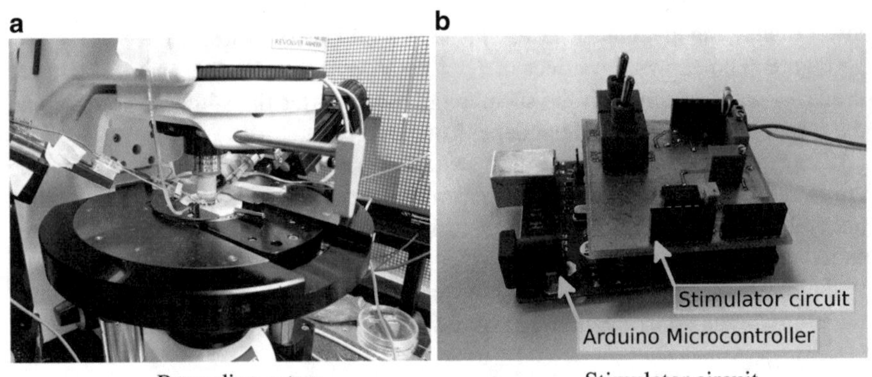

Fig. 4.6 In (**a**) the recording setup is depicted with the stimulation electrode in front and the recording electrode in the back. In (**b**) the realization of the stimulator circuit is shown

using a monophasic cathodic stimulation protocol. The electrode is placed in the extracellular space of the molecular layer in the cerebellum lateral to where the dendritic tree of the Purkinje cells is assumed to be. We aimed to stimulate granule cell axons only to evoke neurotransmitter release and to avoid direct depolarization of the Purkinje cell dendritic tree. Although we cannot exclude that we completely avoided this last possible confounding factor, this setup is sufficient to compare the activation mechanisms of the classical and high frequency stimulation waveforms. An overview of the recording setup is depicted in Fig. 4.6a.

The response to the stimulus is recorded by whole cell patch-clamping Purkinje cells in the voltage-clamp mode using electrodes (the same pipettes as the stimulus electrodes) filled with (in mM): 120 K-Gluconate, 9 KCl, 10 KOH, 3.48 $MgCl_2$, 4 NaCl, 10 HEPES, 4 Na_2ATP, 0.4 Na_3GTP, and 17.5 sucrose, pH 7.25. The membrane voltage is kept at -65 mV and the membrane current is measured using an EPC 10 double patch clamp amplifier (HEKA electronics).

Two different kinds of stimulation are performed and the responses of the Purkinje cell are compared to each other. First of all classical stimulation is applied using a monophasic constant current source. For this purpose a Cygnus Technology SIU90 isolated current source is used. The amplitude of the current is varied to see the effect of stimulation intensity on the response of the Purkinje cell. The stimulation protocol consisted of two consecutive stimulation pulses with a duration of $t_{pulse} = 700\,\mu s$ each and an interpulse interval of 25 ms.

Second, switched-mode stimulation is performed, also using two pulses with $t_{pulse} = 700\,\mu s$ and an interpulse interval of 25 ms. If the Purkinje cell shows a similar response for varying δ during switched-mode as it does for varying amplitude during classical stimulation, it can be concluded that switched-mode stimulation is indeed able to mimic classical stimulation.

4.3.2 Stimulator Design

The circuit used for switched-mode stimulation is depicted in Fig. 4.7. As can be seen a switched-voltage stimulation scheme is applied: transistor M_1 connects the electrode to the stimulation voltage $V_{stim} = -15\,V$, $V_{stim} = -10\,V$, or $V_{stim} = -5\,V$ and is switched with a PWM signal of which the duty cycle δ determines the stimulation intensity.

The PWM signal is generated using the duty cycle generator circuit. Opamps OA_1 and OA_2 generate a triangular signal of which the frequency can be tuned using potentiometer P_1. Subsequently the duty cycle δ is set using potentiometer P_2 at the input of comparator OA_3.

The circuit is controlled using an Arduino Uno microcontroller platform, which also supplies the circuit with a $+5\,V$ supply voltage. The total circuit is isolated from ground by connecting the arduino using the USB of a laptop that is operated from its battery. Capacitor C_2 and clamps D_1 and D_2 are used to level convert the

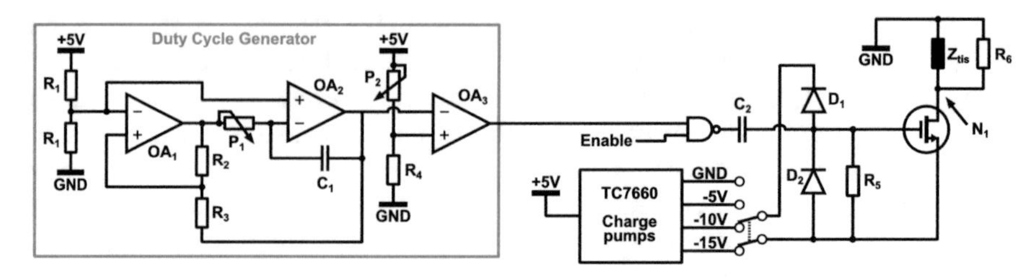

Fig. 4.7 Circuit used to generate a switched-voltage monophasic stimulation protocol

0–5 V logic signal from the duty cycle generator to a V_{stim} to $V_{stim} + 5\,V$ signal to drive the gate of M_1. Resistor $R_6 = 1\,\mathrm{M}\Omega$ is used to discharge the gate of M_1 to V_{stim} in steady state.

Because of the high electrode impedance, any parasitic capacitance connected to node N_1 will prevent the electrode voltage to discharge during the $1 - \delta$ interval of a switching period. This will influence the average voltage over the tissue and the relation between δ and the stimulation intensity. To prevent this effect, resistor $R_5 = 2.7\,\mathrm{k}\Omega$ is placed in parallel with the tissue, which allows the parasitic capacitance to discharge quickly. This resistor does consume power and reduces the power efficiency of the system dramatically. However, the power efficiency is not a design objective for this specific experiment: the only goal is to show the effectiveness of the high frequency stimulation. Without R_5 the stimulation would still be effective, but the electrode voltage would not have the desired switched-mode shape. The whole circuit is implemented on a printed circuit board (PCB), as depicted in Fig. 4.6b.

4.4 Results

In Fig. 4.8a the response of the Purkinje cell is shown for classical constant current stimulation for three different stimulus intensities. First there is a big positive spike corresponding to the stimulation artifact. After a small delay an excitatory postsynaptic current (EPSC) is clearly visible; during this interval the membrane current is decreased due to the opening of the postsynaptic channels of the cell.

After 25 ms the second stimulus arrives and a second EPSC is generated. This EPSC is much bigger due to a process called paired pulse facilitation (PPF): due to the first depolarization the Ca^{2+} concentration in the activated axon terminals is higher when the second pulse arrives, leading to an increased release of neurotransmitter. From the same figure it is also clear that the EPSC becomes stronger for increasing stimulation amplitude.

In Fig. 4.8b the voltage over the stimulation electrode is plotted for various stimulation settings during switched-mode stimulation: both duty cycle δ as well as the supply voltage are varied with a fixed PWM frequency of 100 kHz. Because of the voltage steered character the falling edge of the stimulation pulses is very sharp, while resistance R_5 makes sure that it discharges reasonably fast.

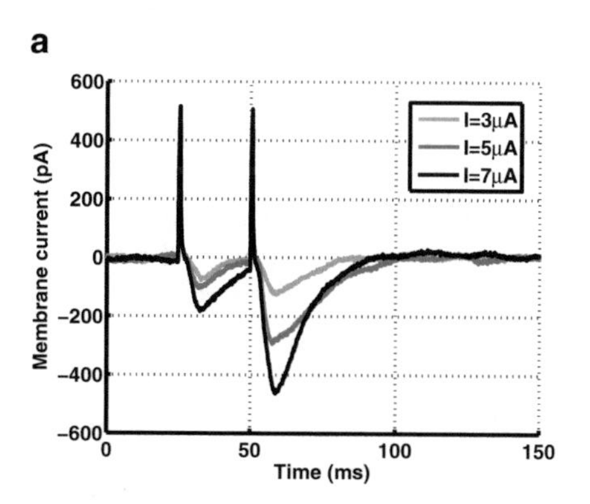

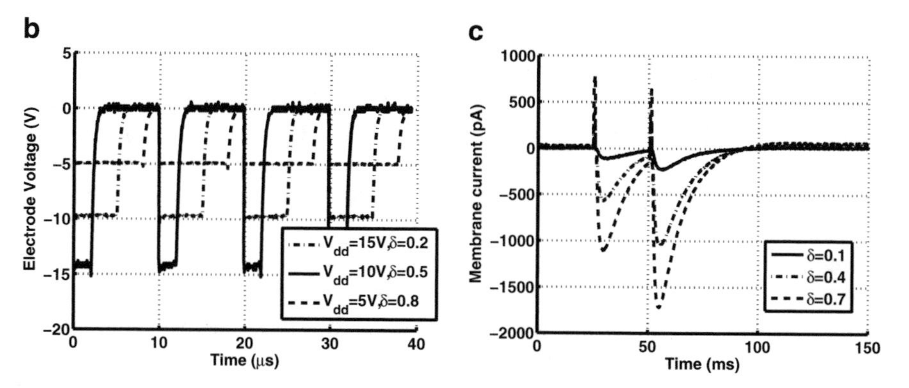

Fig. 4.8 Measurement results from the Purkinje cell during stimulation. In (**a**) patch clamp recordings during classical constant current stimulation are depicted. In (**b**) the electrode voltage during switched-mode stimulation is plotted for various settings of V_{dd} and δ. In (**c**) the response of the neuron to switched-mode stimulation is shown. Both in (**a**) and (**c**) first a positive peak corresponding to the stimulation artifact is seen, after which an EPSC is generated that depends on the stimulation intensity

In Fig. 4.8c the response of the Purkinje cell is shown for switched-mode stimulation. For these plots $V_{dd} = 15\,V$, $t_{pulse} = 700\,\mu\text{s}$, and $f_{stim} = 100\,\text{kHz}$. An EPSC with the same shape as during classical stimulation is the result and also the PPF is clearly visible. It is also seen that by increasing the intensity of the stimulation using δ the EPSC is increased, similar to how it is increased for classical stimulation using the stimulation amplitude. These two points show that the switched-mode stimulation is able to induce similar activity in neural tissue as classical stimulation.

4.5 Discussion

In Fig. 4.9a the absolute value of the minimum in the EPSC $|\min(EPSC)|$ is given as function of the duty cycle δ ($f_{stim} = 100$ kHz, $t_{pulse} = 700\,\mu$s) for the three supply voltages available. Indeed for increasing supply voltage and/or increasing δ the response to the stimulation becomes stronger. This shows that both V_{dd} as well as δ are effective means of adjusting the stimulation intensity.

In Fig. 4.9b the cell is stimulated with $V_{dd} = 5$ V and $t_{pulse} = 700\,\mu$s, but the PWM frequency is varied from 20 kHz up to 100 kHz. As can be seen the stimulation intensity decreases for increasing frequency. This is an unexpected result, based on the simulations using the HH equations in Fig. 4.5d. However, the simulations assumed that all the energy from the voltage source was transferred to Z_{tis}. In reality this is not possible.

In Fig. 4.3b large current peaks can be seen due to the charging of the capacitive component in Z_{tis}. Any resistive component in series with Z_{tis} will reduce V_{tis} (the voltage over Z_{tis}) during such a peak. Examples of these resistances could be a nonzero source impedance, the on resistance of the switch M_1, and the faradaic interface resistance Z_{if} of the electrode. For increasing $f_{stim} = 1/t_s$ the amount of current peaks is increasing, which also increases the losses.

This shows one of the disadvantages of using the switched-mode approach: losses can be expected due to the high frequency components in the stimulation waveform. Therefore, based on the measurement results, it can be concluded that switched-mode stimulation can lead to the same activation as classical stimulation, but care has to be taken to minimize additional losses that may arise due to the high frequency operation.

This conclusion confirms the electrophysiological feasibility for the design of stimulators that employ a high frequency output. These systems can improve on

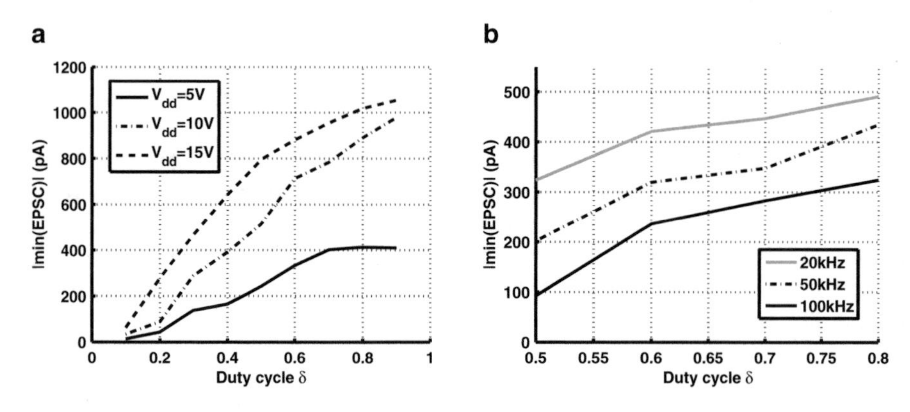

Fig. 4.9 In (**a**) the absolute value of the minimum EPSC is plotted as a function of δ for various settings of V_{dd}. In (**b**) the absolute value of the minimum EPSC is plotted for several PWM frequencies

important aspects such as power efficiency [4] and size [3] of the stimulator. A trade-off needs to be made between the advantages that switched-mode operation can offer versus the additional losses.

This chapter didn't address the consequences for tissue damage due to the use of the switched-mode approach. Most of the studies analyzing tissue damage [15, 16] use a classic stimulation scheme only and therefore it is not known how their results translate to switched-mode operation. Furthermore, the losses due to the high frequency operation are not quantified, since the stimulator circuit that was used did not allow for that. It would be required to compare the EPSC with the total amount of charge injected in the tissue (and not R_5) during the stimulation pulse. Further investigation is needed to address these issues.

4.6 Conclusions

In this chapter a theoretical analysis and in vitro experiments were used to verify the efficacy of high frequency switched-mode stimulation. Using modeling that included the dynamic properties of both the tissue material as well as the axon membrane it was found that high frequency stimulation signals can recruit neurons in a similar fashion as classical constant current stimulation.

The response of Purkinje cells due to stimulation in the molecular layer was measured for both classical and switched-mode stimulation. The measurements confirmed the modeling in showing that switched-mode stimulation can induce neuronal activation and that both the duty cycle δ and the stimulation voltage V_{stim} are effective ways to control the intensity of the stimulation. This shows that from an electrophysiological point of view, it is feasible to use high frequency stimulation, which paves the way for the design of switched-mode stimulator circuits. In Chap. 7 of this book a neural stimulator is presented that takes advantage of this approach. Care has to be taken to avoid losses in the stimulation system that arise due to the use of a high frequency stimulation signal.

References

1. Sahin, M., Tie, Y.: Non-rectangular waveforms for neural stimulation with practical electrodes. J. Neural Eng. **4**(3), 227–233 (2007)
2. Wongsarnpigoon, A., Grill, W.M.: Energy-efficient waveform shapes for neural stimulation revealed with a genetic algorithm. J. Neural Eng. **7**(4), 0460009 (2010)
3. Liu, X., Demosthenous, A., Donaldson, N.: An integrated implantable stimulator that is fail-safe without off-chip blocking-capacitors. IEEE Trans. Biomed. Circuits Syst. **2**(3), 231–244 (2008)
4. Arfin, S.K., Sarpeshkar, R.: An energy-efficient, adiabatic electrode stimulator with inductive energy recycling and feedback current regulation. IEEE Trans. Biomed. Circuits Syst. **6**(1), 1–14 (2012)

5. van Dongen, M.N., Hoebeek, F.E., Koekkoek, S.K.E., De Zeeuw, C.I., Serdijn, W.A.: High frequency switched-mode stimulation can evoke postsynaptic responses in cerebellar principal neurons. Front. Neuroengineering **8**(2) (2015). http://journal.frontiersin.org/article/10.3389/fneng.2015.00002/abstract
6. Merrill, D.R., Bikson, M., Jefferys, J.G.R.: Electrical stimulation of excitable tissue – design of efficacious and safe protocols. J. Neurosci. Methods **141**, 171–198 (2005)
7. Gabriel, S., Lau, R.W., Gabriel, C.: The dielectric properties of biological tissues III: parametric models for the dielectric spectrum of tissues. Phys. Med. Biol. **41**(11), 2271–2293 (1996)
8. Warman, E.N., Grill, W.M., Durand, D.: Modeling the effects of electric fields on nerve fibers: determination of excitation thresholds. IEEE Trans. Biomed. Eng. **39**(12), 1244–1254 (1992)
9. Bosseti, C.A., Birdno, M.J., Grill, W.M.: Analysis of the quasi-static approximation for calculating potentials generated by neural stimulation. J. Neural Eng. **5**(1), 44–53 (2008)
10. Tai, C., de Groat, W.C., Roppolo, J.R.: Simulation analysis of conduction block in unmyelinated axons induced by high-frequency biphasic electrical currents. IEEE Trans. Biomed. Eng. **52**(7), 1323–1332 (2005)
11. Somogyi, P., Hamori, J.: A quantitative electron microscopic study of the purkinje cell axon initial segment. Neuroscience **1**(5), 361–365 (1976)
12. Hodgkin, A.L., Huxley, A.F.: A quantitative description of membrane current and its application to conduction and excitation in nerve. J. Physiol. **117**(4), 500–544 (1952)
13. Rattay, F.: Analysis of models for extracellular fiber stimulation. IEEE Trans. Biomed. Eng. **36**(7), 676–682 (1989)
14. Gao, Z., Todorov, B., Barrett, C.F., van Dorp, S., Ferrari, M.D., van den Maagdenberg, A.M.J.M., De Zeeuw, C.I., Hoebeek, F.E.: Cerebellar ataxia by enhanced $Ca_V2.1$ currents Is alleviated by Ca^{2+}-dependent K^+-channel activators in cacna1a^{S218L} mutant mice. J. Neurosci. **32**(44), 15533–15546 (2013)
15. Shannon, R.V.: A model of safe levels for electrical stimulation. IEEE Trans. Biomed. Eng. **39**(4), 424–426 (1992)
16. Butterwick, A., Vankov, A., Huie, P., Freyvert, Y., Palanker, D.: Tissue damage by pulsed electrical stimulation. IEEE Trans. Biomed. Eng. **54**(12), 2261–2267 (2007)

Part II
Electrical Design of Neural Stimulators

This part discusses the electrical design of neural stimulators. It uses several concepts from the previous part about safe and efficient stimulation and implements these on circuit level. Two stimulator systems with completely different applications are discussed and their design and validation is shown.

Chapter 5
System Design of Neural Stimulators

Abstract This chapter discusses several system aspects for the design of neural stimulator circuits and provides a framework to compare such designs in general. Throughout the chapter, a comparison is made between the neural stimulator designs that are to be discussed in more detail in Chaps. 6 and 7. Both designs have been created with a different application in mind and this chapter discusses the consequences.

The system from Chap. 6 is designed to be used for animal experiments and aims to provide maximum flexibility in the stimulation settings that can be used. The system is not intended to be implantable or wearable. The system from Chap. 7 is designed for implantable applications and benefits from the stimulation strategy that was outlined in Chap. 4.

In the first section of this chapter some system level design aspects are discussed such as the location of the system, the electrode configuration, the stimulation waveform, and the charge cancellation. In the second section a closer look is given to the electrical implementation of the stimulator system.

5.1 System Properties of Neural Stimulators

In this section various general system properties of neural stimulators are discussed and it is shown which of these properties are chosen for the systems in the subsequent chapters.

5.1.1 Location of the System

Transcutaneous systems are completely outside of the body and hence do not include any implanted part. Ancient forms of electrical stimulation, such as the use of electrical fish, can be considered transcutaneous. Today transcutaneous electrical nerve stimulation (TENS) is a therapeutic procedure for pain suppression [1, 2],

M. van Dongen, W. Serdijn, *Design of Efficient and Safe Neural Stimulators*,
Analog Circuits and Signal Processing, DOI 10.1007/978-3-319-28131-5_5

although its efficacy remains unclear. One obvious advantage of TENS is that it is non-invasive, while a disadvantage is poor selectivity of the target stimulation area. System requirements are generally more relaxed due to the fact that nothing is implanted.

Percutaneous (through-the-skin) systems use implanted electrodes to improve the selectivity of the stimulation. The electrodes connect with wires through the skin to an external stimulator unit. In Chap. 6 this type of stimulator is designed for use in animal studies for tinnitus treatment.

Implantable systems do not have external components and are fully implanted. This puts strict design requirements on the system in terms of size and bio-compatibility. Chapter 7 describes the design of a power efficient stimulator front-end intended for implantable solutions.

5.1.2 Electrode Configuration

A general overview of a multichannel stimulator system is given in Fig. 5.1a: N stimulation sources connect to M electrodes. Depending on the application, the electrodes may be placed in monopolar, bipolar, or (if $M > 2$) multipolar configurations. Multipolar configurations allow for field steering techniques, as was illustrated in Fig. 2.6.

A single channel stimulator ($N = 1$) consists of one stimulation source that can be connected to two or more electrodes. This means that it is only possible to provide one stimulation pattern simultaneously on the electrodes. The system described in Chap. 6 is a single channel stimulator system that is connected to 8 electrodes.

Multichannel systems allow for more complicated configurations. In Fig. 5.1b a common topology used for systems with a high number of electrodes, such as retinal or cochlear implants, is shown. Each stimulation channel is operated in a monopolar

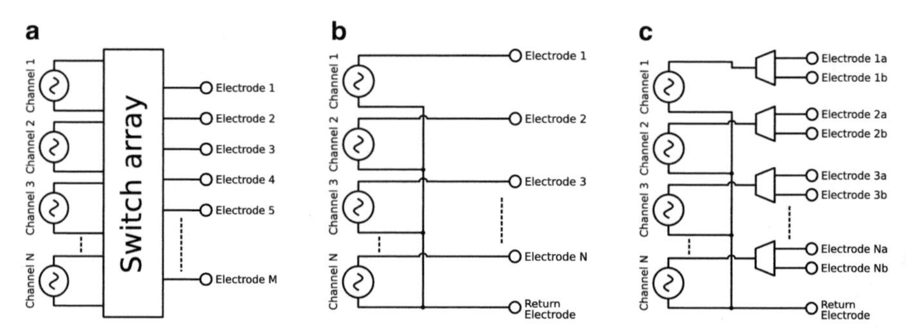

Fig. 5.1 Several possibilities of connecting a multichannel stimulator system to multiple electrodes. In (**a**) a general system with N channels connecting to M electrodes is depicted that offers maximum flexibility. In (**b**) and (**c**) a common topology for cochlear and retinal implants is shown in which each channel is fixed to one or a few electrodes

fashion, which means that each channel connects to one electrode, while the return current flows through a common return electrode. Each channel can be stimulated simultaneously and independently, but is connected to one specific electrode. The number of channels can be increased further by incorporating a demultiplexer at the output as depicted in Fig. 5.1c. In [3] the design of a 256 channel system with 1024 electrodes is proposed using this topology.

The configurations of Figs. 5.1b and c impose limits on the flexibility of the system: each channel can stimulate only one electrode; it is not possible to stimulate other electrodes with that channel and it is also not possible to involve multiple channels with the same electrode. These features are important when advanced field steering techniques are required. The system in Chap. 7 has 8 independent channels that can connect to any of the 16 electrodes at the output. This system enables endless possibilities for electrode configurations and current steering techniques.

5.1.3 Stimulation Waveform

As shown in Sect. 2.2.3 the recruitment of neurons depends on the intensity and duration of the stimulation waveform. However, the stimulation waveform can be generated in many different ways and certain choices for the waveform will have important implications for efficacy, safety, and efficiency.

5.1.3.1 Output Quantity

There are several reasons why most classical stimulators use constant current stimulation. First of all, as was seen in Chap. 2 (e.g., in Eq. 2.13), the stimulation current controls the potential in the tissue and therefore determines the intensity of the stimulation. By using current based stimulation, the stimulation intensity is directly controllable. Another related reason is that using this approach the amount of charge injected is controlled more easy.

Besides current controlled stimulation, it is also possible to use voltage controlled stimulation [4]. In this case the current through the load and hence the stimulation intensity and the injected charge depend on the tissue impedance. This type of stimulator often includes a circuit to keep track of the injected charge. Charge steered stimulators based on switched-capacitor techniques have also been proposed [5].

It is important to realize that from a stimulation point of view it is not required to have a well defined stimulation quantity. The amplitude of the stimulation signal is determined by the response of the subject and needs to be determined empirically or in a closed-loop fashion by monitoring the (neural) response and adjusting the intensity of the stimulation accordingly. Therefore the absolute value of the stimulation signal is not important.

The stimulator system proposed in Chap. 6 can be both current or voltage controlled, but it is implemented as a current controlled system. The system implemented in Chap. 7 uses a novel stimulation principle that could be labeled as flux controlled. Both implementations put only limited emphasis on creating accurate output quantities.

5.1.3.2 Arbitrary Waveforms

Classical stimulation schemes use a constant stimulation intensity. Several studies however have shown that other waveforms can lead to more effective stimulation. The stimulation waveform can benefit from the dynamic properties of the membrane as described by the Hodgkin–Huxley equations. In [6] it was shown that Gaussian stimulation waveforms need less energy to obtain the same neural recruitment as compared to rectangular waveforms. Furthermore, the waveform also has influence on the charge injection capacity of the electrodes. In [7] an algorithm was presented to find the optimal stimulation waveform from an energy efficiency point of view and again Gaussian waveforms were found.

The use of non-constant stimulation waveforms makes some aspects of the system design more complicated, such as the charge balancing scheme. The stimulator designed in Chap. 6 offers arbitrary waveforms, while still offering charge balanced stimulation. The system in Chap. 7 uses a constant stimulation intensity.

5.1.3.3 Monophasic vs Biphasic

Another characteristic of the stimulation waveform is the monophasic or biphasic nature. In [8] it was shown that monophasic stimulation is more efficacious than biphasic stimulation, due to the fact that the second stimulation phase can (partially) cancel the effect of the first stimulus.

However, it is not always possible to use monophasic stimulation. As seen in Chap. 3, only non-polarizable electrodes such as Ag-AgCl electrodes can use monophasic stimulation waveforms, because there is no risk of charge accumulation at the interface. Polarizable electrodes need biphasic stimulation waveforms that discharge the interface capacitor C_{dl} as shown in Fig. 5.2. Both Chaps. 6 and 7 discuss stimulators that deliver biphasic stimulation pulses.

Constant current stimulators often implement biphasic stimulation using a symmetrical scheme. In this way the charge in both stimulation phases is controlled by using the same pulsewidth. Since the system in Chap. 6 allows for arbitrary waveforms, it also allows for asymmetrical biphasic stimulation schemes.

5.1.3.4 Stimulation Parameters

The *amplitude* and *pulse width* of the stimulation pulse determine the stimulation intensity, which is summarized in the strength-duration curve as discussed in Chap. 2. Stimulators need to be able to adjust either one of them or both in order to control the stimulation intensity.

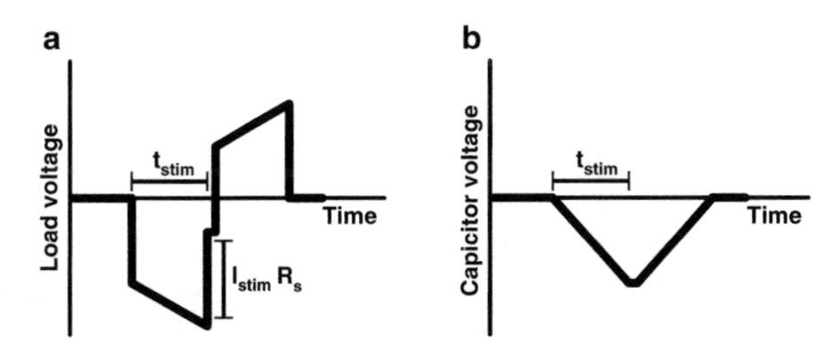

Fig. 5.2 Illustration of the voltage waveforms resulting from a constant current biphasic stimulator connected to a series R_s-C_{dl} model. In (**a**) the voltage over the complete load is sketched, while in (**b**) the voltage over the interface capacitance C_{dl} is shown. Charge cancellation schemes aim to prevent charge accumulation at C_{dl} over multiple stimulation cycles

When a biphasic stimulation pulse is used, the *interpulse delay* is the time between the first stimulation pulse and the second. The longer this time is, the longer the membrane voltage of the target tissue stays depolarized or hyperpolarized before the charge cancellation pulse reverses it. This gives more time for the Hodgkin–Huxley dynamics to 'react' and therefore the stimulation becomes more efficacious [8, 9]. On the other hand, the electrode–tissue interface capacitance C_{dl} will be charged longer (see Fig. 5.2), which can cause tissue damage. In both systems from Chaps. 6 and 7 the interpulse delay is adjustable.

Stimulation pulses can be triggered in a variety of ways. In case of *single shot stimulation* a single stimulation pulse is triggered when the stimulation is started, for example, by sending a stimulation command to the system. This type of stimulation is typically used in closed-loop stimulator systems, for example, in epileptic seizure suppression [10, 11]. In case of *tonic stimulation*, the stimulation pattern is repeated in a synchronous fashion with a certain fixed repetition frequency. This type of stimulation is the most common and is used in a wide variety of applications: in TENS, PNS, VNS, and DBS. Repetition frequencies range from a few Hz up to 1 kHz [12, 13]. In case of *burst stimulation*, a number of stimulation pulses are injected fast after each other with a certain frequency. Clinical experiments have shown that this type of stimulation can be more effective than tonic stimulation [14].

5.1.4 Charge Cancellation Schemes

As explained in Chap. 3, it is important for polarizable electrodes not to accumulate any charge over the electrode–tissue interface. Referring to Fig. 5.2, it is important that the voltage over the capacitor after the stimulation cycle is as close to zero as possible. Various charge cancellation techniques have been proposed [15], several of which have been implemented in the stimulator systems discussed in Chaps. 6 and 7.

5.1.4.1 Electrode Shortening (Passive Discharge)

Shortening the electrodes will remove the charge that is left at the interface by passively discharging it. The discharge rate will depend on the time constant set by the interface capacitance C_{dl} itself and the tissue resistance R_s.

Passive discharging is not suitable for high stimulation rates and/or high impedance electrodes, because the interface will not have the chance to fully discharge. This will cause charge build-up over multiple stimulation cycles. However, it is a useful method to combine with other charge cancellation techniques, which bring the interface voltage sufficiently close to zero already. This strategy is therefore applied in both systems of Chaps. 6 and 7.

5.1.4.2 Coupling Capacitors

The easiest way to ensure that no DC is passed through the load is by connecting a coupling capacitor in series with the tissue: as shown in Chap. 3 the capacitor will ensure that the average voltage over the electrode–tissue interface is zero. An additional advantage of these capacitors is that they also prevent DC currents to flow through the electrodes in case of a device failure.

However, Chap. 3 also showed the disadvantages of using coupling capacitors. It was shown that they do not improve the charge balancing as compared to passive discharge without coupling capacitors. Furthermore, a coupling capacitor introduces a shift in the equilibrium voltage of the electrode–tissue interface. Therefore care has to be taken not to introduce too high DC offsets when coupling capacitors are used.

Another drawback of these capacitors is the space they require. The value of these capacitors should be large enough in order to prevent a significant voltage over them during a regular stimulation cycle and their values are therefore often in the range of hundreds of nanofarads [16] up to microfarads [17]. These capacitors are therefore often realized using external discrete components.

5.1.4.3 Charge/Current Monitoring

One way to control the injected charge is by measuring the stimulation current. In some cases this measure is used to accurately match the stimulation current during both stimulation phases [18–20]: push/pull matching. By also accurately controlling the stimulation time, the total charge is balanced, depending on the accuracy of the current matching and timing.

Current monitoring is also often used in voltage steered systems [4] or in systems in which the injected charge is not easily controlled, such as arbitrary waveform stimulators [21]. The system described in Chap. 6 uses this technique: it measures the stimulation current and integrates this to obtain a value for the injected charge.

The accuracy of the charge balancing for systems using this approach is limited by the accuracy of the current measurement. Therefore, this technique often needs to be combined with other charge cancellation methodologies to ensure safe operation.

5.1.4.4 Pulse Insertion

All previous charge balancing techniques do not include feedback in the charge balancing scheme. Pulse insertion introduces feedback by measuring the tissue voltage after a stimulation cycle is finished. When the voltage over the interface is not within a safe margin, this indicates that too much charge is left. In that case, a number of charge packets are injected by means of a short current pulse until the interface voltage returns to a safe value [22].

Advantage of this technique is that the interface is guaranteed to be brought back within a safe window. Care has to be taken though that the pulses inserted will not lead to undesired stimulation artifacts in the target tissue. This technique is implemented in the system of Chap. 7.

5.1.4.5 Offset Injection

Offset injection [15] is another charge balancing technique that uses feedback. Based on the remaining interface voltage after a stimulation cycle, an offset is added to the next stimulation cycle. By continuously regulating this offset, the remaining voltage over the interface capacitor is brought back to zero volts.

Advantage of this method is that there are no artifacts that may influence the recruitment of target neural tissue. Drawback is that the offset changes the total amount of injected charge, which will also influence the volume in the tissue which will be recruited by the stimulation.

5.1.4.6 *IR*-Drop Measurement

The methodology introduced in Chap. 3 uses a feedforward mechanism based on the value of the $I_{\text{stim}}R_s$-drop indicated in Fig. 5.2. Advantage of this methodology is that the total amount of charge injected during stimulation is not affected and that also no artifacts due to pulse insertion are introduced.

However, it was also shown in Chap. 3 that this method depends on the validity of the simple series R_s-C_{dl} model that is used for the electrodes and the tissue. Moreover it was shown that in some cases this model is no longer valid, which will cause the remaining voltage after a stimulation cycle to be not equal to zero.

5.2 System Implementation Aspects

In this section several general aspects for the circuit implementation of the output stage are discussed. A closer look is given to the power efficiency of the output stage, as well as on the mechanism to generate the biphasic stimulation scheme.

5.2.1 *Power Efficiency of Neural Stimulators*

For battery operated stimulators, the power efficiency is an important factor in the lifetime of the device, along with the required battery size. Over the years several output stage configurations have been proposed in order to improve the power efficiency. This section reviews them shortly.

- *Linear operation*
  Figure 5.3a shows the most straightforward implementation of a current source based output stage. The stimulation current is injected in the tissue by controlling the switches using non-overlapping signals. The stimulation current I_{stim} is regulated by a current source that is supplied from a fixed supply voltage V_{dd}. Most early current based stimulator systems were based on this topology [18, 23, 24].

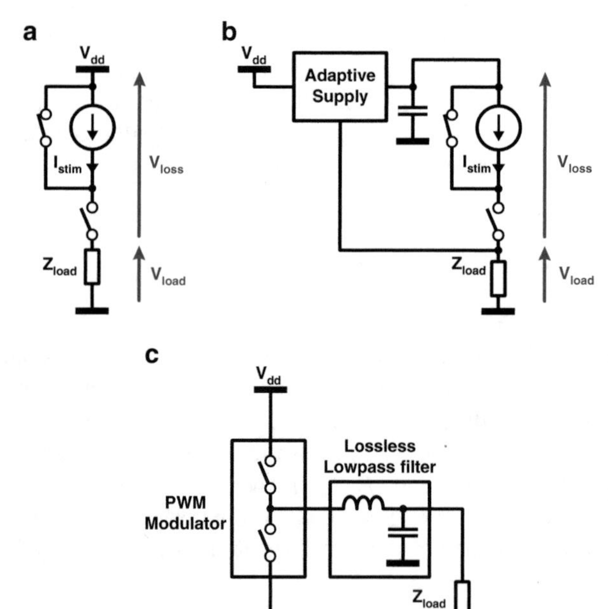

Fig. 5.3 Three possible output stage topologies (current steered) with varying power efficiencies. In (**a**) a linear output stage is depicted where the power efficiency is limited due to the voltage drop V_{loss} over the current driver. In (**b**) V_{loss} is minimized by adapting the supply voltage to the load voltage (class G/H operation). In (**c**) a switched-mode topology is depicted in which the power efficiency is determined by the switching and conduction losses in the PWM modulator

The efficiency of these systems is limited due to the voltage drop over the driver, which depends on the load conditions: $V_{loss} = V_{dd} - I_{stim}Z_{load}$. The power efficiency of the current source is now found as $\eta = V_{load}/V_{loss}$. Since Z_{load} and I_{stim} can vary over many orders of magnitude for a single application, V_{dd} needs to be chosen relatively high in order to accommodate for the worst case scenario. Therefore these systems suffer from limited power efficiency for typical operating conditions, in which $V_{load} \ll V_{loss}$.

- *Adaptive power supply (class G/H)*
  To minimize the value of V_{loss}, the supply voltage V_{dd} can be adapted to the load voltage V_{load} by means of an adaptive supply. This is more or less equivalent to class G/H operation and is schematically depicted in Fig. 5.3b, in which the switches are again controlled using non-overlapping signals. Many state-of-the-art neural stimulator designs that focus on high power efficiency employ this type of topology [3, 25–27] and use a so-called compliance monitor to adapt V_{dd}.
  The efficiency is determined by the minimum required value of V_{loss} to make the current source work, the efficiency of the adaptive supply generator, and the number of supply levels that is available. In Chap. 7 the power efficiency of this type of system is analyzed quantitatively.
- *Switched-mode operation (class D)*
  In [4] a switched-mode stimulator system was proposed for which the operation is comparable to class D operation. An overview of the topology is depicted in Fig. 5.3c: by low-pass filtering a Pulse Width Modulated (PWM) square wave signal the required stimulation signal is transferred to the load, without suffering from effects such as V_{loss}. Instead the power efficiency is mainly determined by the switching and conduction losses in the PWM modulator.
  This system needs additional passive components for the lossless output filter. Furthermore, the output quantity is voltage, which requires additional current monitoring circuitry if the desired output quantity is current [4].

The system designed in Chap. 6 uses a linear output stage, since power efficiency is not an important design criterion for percutaneous systems. However, the principles outlined in this chapter are still suitable to be combined with an adaptive supply system. The system described in Chap. 7 uses an alternative power efficient output topology, which is based on switched-mode operation. It reduces the number of external components and is not voltage controlled.

5.2.2 Bidirectional Stimulation

In order to generate a bidirectional stimulation pulse, two principles are possible, as illustrated in Fig. 5.4. In Fig. 5.4a two separate supplies are used for each stimulation phase. In most cases, the tissue is grounded on one side and two symmetrical supplies are used for the current sources, but in some cases the tissue is connected to a midrail reference, which allows for positive supply voltages exclusively [4]. These implementations are equivalent from a power efficiency perspective.

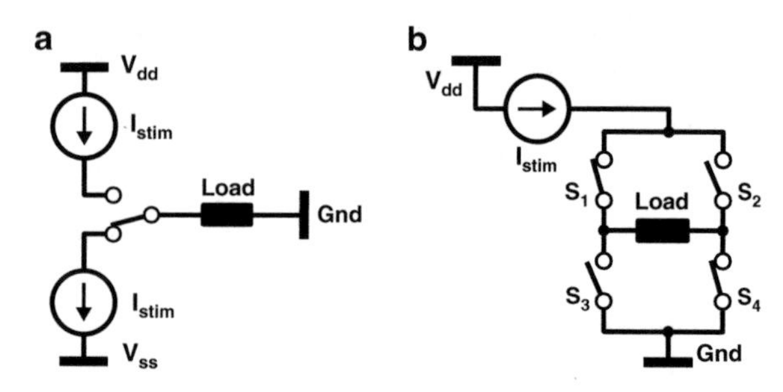

Fig. 5.4 Two possible implementations of a biphasic stimulator. In (**a**) two supplies are used, while in (**b**) an H-bridge is employed. In this figure current sources are used, but it is possible to replace them with other types as well (voltage, charge, etc.)

Table 5.1 Overview of the properties of the stimulator systems designed in Chaps. 6 and 7

	Chapter 6	Chapter 7
Invasiveness	Percutaneous	Implantable
Electrode configuration	Mono- or bipolar	Arbitrary
Stimulation quantity	Current	Flux
Adjustable amplitude	Yes	Yes
Adjustable pulsewidth	Yes	Yes
Adjustable interpulse delay	Yes	Yes
Single shot stimulation	Yes	Yes
Tonic stimulation	Yes	Yes
Burst Stimulation	Yes	No
Number of channels	1	8
Number of electrodes	8	16
Waveform	Arbitrary	Current spikes
Monophasic/Biphasic	Biphasic (asymmetrical)	Biphasic (symmetrical)
Charge cancellation	Current monitoring	Pulse insertion
Output topology	Linear	Switched-mode

In Fig. 5.4b an H-bridge technique is implemented that uses a single current source only. During the first stimulation phase only switches S_1 and S_4 are closed, while during the second phase only S_2 and S_3 are used. Advantage of the H-bridge approach is that only a single supply and stimulation source are needed. Drawback is the more complicated switching scheme and the fact that upon reversal of the stimulation direction, the charged interface capacitor introduces a negative voltage (below ground), that might cause leakage through the substrate.

5.3 Conclusions

In this chapter an overview was given of several important system properties of neural stimulators. In Table 5.1 an overview is given of how these properties are implemented in the systems to be discussed in Chaps. 6 and 7. The system in Chap. 6 focuses on flexibility in stimulation waveforms, while power efficiency is of smaller concern for an external percutaneous system. The system in Chap. 7 focuses on a power efficient implantable solution that can be configured for many electrode configurations.

References

1. Marchand, S., Charest, J., Li, J., Chenard, J.R., Lavignolle, B., Laurencelle, L.: Is TENS purely a placebo effect? A controlled study on chronic low back pain. Pain **54**(1), 99–106 (1993)
2. Deyo, R.A., Walsh, N.E., Martin, D.C., Schoenfeld, L.S., Ramamurthy, S.: A controlled trial of transcutaneous electrical nerve stimulation (TENS) and exercise for chronic low back pain. N. Engl. J. Med. **322**(23), 1627–1634 (1990)
3. Noorsal, E., Sooksood, K., Xu, H., Hornig, R., Becker, J., Ortmanns, M.: A neural stimulator frontend with high-voltage compliance and programmable pulse shape for epiretinal implants. IEEE J. Solid State Circuits **47**(1), 244–256 (2012)
4. Arfin, S.K., Sarpeshkar, R.: An energy-efficient, adiabatic electrode stimulator with inductive energy recycling and feedback current regulation. IEEE Trans. Biomed. Circuits Syst. **6**(1), 1–14 (2012)
5. Ghovanloo, M.: Switched-capacitor based implantable low-power wireless microstimulating systems. Proceedings of the 2006 IEEE International Symposium on Circuits and Systems (2006)
6. Sahin, M., Tie, Y.: Non-rectangular waveforms for neural stimulation with practical electrodes. J. Neural Eng. **4**(3), 227–233 (2007)
7. Wongsarnpigoon, A., Grill, W.M.: Energy-efficient waveform shapes for neural stimulation revealed with a genetic algorithm. J. Neural Eng. **7**(4), 046009 (2010)
8. Merrill, D.R., Bikson, M., Jefferys, J.G.R.: Electrical stimulation of excitable tissue – design of efficacious and safe protocols. J. Neurosci. Methods **141**, 171–198 (2005)
9. Hofmann, L., Ebert, M., Tass, P.A., Hauptmann, C.: Modified pulse shapes for effective neural stimulation. Front. Neuroengineering **4**, 9 (2011)
10. Berényi, A., Belluscio, M., Mao, D., Buzsáki, G.: Closed-loop control of epilepsy by transcranial electrical stimulation. Science **337**, 735 (2012)
11. Paz, J.T., Davidson, T.J., Frechette, E.S., Delord, B., Parada, I., Peng, K., Diesseroth, K. Huguenard, J.R.: Closed-loop optogenetic control of thalamus as a tool for interrupting seizures after cortical injury. Nat. Neurosci. **16**(1), 64–70 (2013)
12. Chesterton, L.S., Foster, N.E., Wright, C.C., Baxter, G.D., Barlas, P.: Effects of TENS frequency, intensity and stimulation site parameter manipulation on pressure pain thresholds in healthy human subjects. Pain **106**(1-2), 73–80 (2003)
13. Kuncel, A.M., Grill, W.M.: Selection of stimulus parameters for deep brain stimulation. Clin. Neurophysiol. **115**(11), 2431–2441 (2004)
14. De Ridder, D., Vanneste, S., Loo, E. van der Plazier, M., Menovsky, T., van de Heyning, P.: Burst stimulation of the auditory cortex: a new form of neurostimulation for noise-like tinnitus suppression. J. Neurosurg. **112**(6), 1289–1294 (2010)
15. Sooksood, K, Stieglitz, T., Ortmanns, M.: An active approach for charge balancing in functional electrical stimulation. IEEE Trans. Biomed. Circuits Syst. **4**(3), 162–170 (2010)
16. Constandinou, T.G., Georgiou, J., Toumazou, C.: A partial-current-steering biphasic stimulation driver for vestibular prostheses. IEEE Trans. Biomed. Circuits Syst. **2**(2), 106–113 (2008)
17. Techer, J.D., Bernard, S., Bertrand, Y., Cathebras, G., Guiraud, D.: New implantable stimulator for the FES of paralyzed muscles. Proceeding of the 30th European Solid-State Circuits Conference ESSCIRC, pp. 455–458 (2004)
18. Site, J.J., Sarpeshkar, R.: A low-power blocking-capacitor-free charge-balanced electrode-stimulator chip with less than 6 nA DC error for 1-mA full-scale stimulation. IEEE Trans. Biomed. Circuits Syst. **1**(3), 172–183 (2007)

19. Lee, E., Lam, A.: A matching technique for biphasic stimulation pulse. IEEE International Symposium on Circuits and Systems, pp. 817–820 (2007)
20. Xiang, F., Wills, J., Granacki, J., LaCoss, J., Arakelian, A., Weiland, J.: Novel charge-metering stimulus amplifier for biomimetic implantable prosthesis. IEEE International Symposium on Circuits and Systems, pp. 569–572 (2007)
21. van Dongen, M.N., Serdijn, W.A.: Design of a versatile voltage based output stage for implantable neural stimulators. IEEE First Latin American Symposium on Circuits and Systems (2010)
22. Ortmanns, M., Rocke, A., Gehrke, M., Teidtke, H.J.: A 232-channel epiretinal stimulator ASIC. IEEE J. Solid-State Circuits **42**(12), 2946–2959 (2007)
23. Bhatti, P.T., Wise, K.D.: A 32-site 4-channel high-density electrode array for a cochlear prosthesis. IEEE J. Solid-State Circuits **41**(12), 2965–2973 (2006)
24. Coulombe, J., Sawan, M., Gervais, J.F.: A highly flexible system for microstimulation of the visual cortex: design and implementation. IEEE Trans. Biomed. Circuits Syst. **1**(4), 258–269 (2007)
25. Sooksood, K., Noorsal, E., Bihr, U., Ortmanns, M.: Recent advances in power efficient output stage for high density implantable stimulators. 2012 IEEE Annual International Conference of the Engineering in Medicine and Biology Society (EMBS), pp. 855–858 (2012)
26. Williams, I., Constandinou, T.G.: An energy-efficient, dynamic voltage scaling neural stimulator for a proprioceptive prosthesis. IEEE Trans. Biomed. Circuits Syst. **7**(2), 129–139 (2013)
27. Lee, H.N., Park, H., Ghovanloo, M.: A power-efficient wireless system with adaptive supply control for deep brain stimulation. IEEE J. Solid-State Circuits **48**(9), 2203–2216 (2012)

Chapter 6
Design of an Arbitrary Waveform Charge Balanced Stimulator

Abstract This chapter discusses the design of an arbitrary waveform, charge balanced biphasic stimulator. The philosophy behind the system is to give the user full flexibility in the choice for the stimulation waveform, while the safety is ensured by implementing a charge balance mechanism. As discussed in Chap. 5 the stimulation waveform can benefit from the complex dynamics of the axon membrane in order to induce recruitment in a more efficient way. Furthermore, burst stimulation can be considered to be a special stimulation waveform as well, which has also shown to have advantages.

In the first section the general system design is discussed, focusing on the current monitoring technique that is used for charge balancing. In Sect. 6.2 the circuit design for an IC implementation of the system is discussed and the simulation results are presented.

In Sect. 6.3 a discrete realization of the proposed technique is discussed and the measurement results are presented. It was chosen to realize the implementation in discrete form as opposed to an IC because of the application of the system in a percutaneous *in vivo* experiment that is discussed in Sect. 6.4. The discrete system allows for more rapid prototyping and more flexibility. Furthermore the percutaneous nature of the experiment greatly relaxes the requirements on power consumption and size of the device.

6.1 System Design

In traditional constant current stimulator systems, the injected charge is controlled by considering $Q = I_{stim}t_{stim}$ with I_{stim} the stimulation current and t_{stim} the stimulation duration. It is therefore easy to achieve approximate charge balancing by using anodic and cathodic phases with identical I_{stim} and t_{stim} (symmetrical biphasic stimulation).

For the system designed in this chapter the stimulation waveforms are arbitrary and hence not of constant amplitude. However, it is required to match the charge

M. van Dongen, W. Serdijn, *Design of Efficient and Safe Neural Stimulators*,
Analog Circuits and Signal Processing, DOI 10.1007/978-3-319-28131-5_6

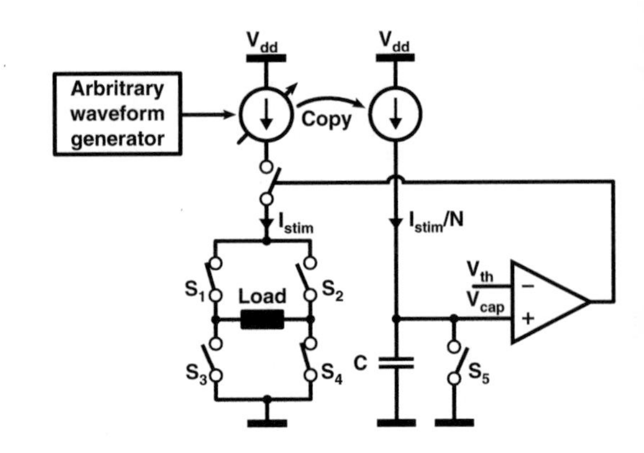

Fig. 6.1 System design of the arbitrary waveform stimulator. The stimulation current I_{stim} is copied as I_{stim}/N and is integrated on C to obtain a measure for the charge in the form of V_{cap}

contents of both stimulation phases. One way to achieve this is by incorporating a current sensor in the stimulation path, from which the charge follows as $Q = \int I_{stim}(t)dt$.

One way to do this is by using a sense resistor in the current path. In [1] a 200 Ω resistor is placed in series with the stimulation source and using a switched-capacitor integrator the total charge during a stimulation phase is determined by monitoring the differential voltage over this resistor. Drawbacks of this approach include the losses introduced by the resistor (the system assumes an electrode with $R_s = 5\,\mathrm{k}\Omega$ and hence the sense resistor reduces the efficiency by 4 %). Furthermore, the common mode voltage over the sense resistor changes very abruptly during the onset of a stimulation phase, which puts strict requirements on the integrator.

In the proposed system it is chosen to make an accurate and scaled copy of I_{stim} that can be processed further in a current integrator [2], as is schematically shown in Fig. 6.1. The current I_{stim}, that is controlled by an arbitrary waveform generator, is used to stimulate the tissue, while a scaled copy I_{stim}/N is sent to an integrator (implemented using a capacitor C) to determine the total charge of the stimulation pulse.

The system can work in two different modes. In the first mode, a predefined amount of charge is selected by the user by means of V_{th}. During both the first stimulation phase (S_1 and S_4 are closed) as well as during the second phase (S_2 and S_3 are closed), the stimulation is stopped when $V_{cap} = V_{th}$, meaning the required amount of charge is injected into the tissue. After both stimulation phases the integrator is reset by closing switch S_5. Drawback of this mode is that the pulsewidth of the stimulation pulse is not directly controlled by the user.

In the second mode, a predefined stimulation pulse with a certain pulsewidth is injected, while $V_{th} = V_{dd}$, which prevents the stimulation from being stopped by reaching a charge limit. After this first pulse, the value of $V_{th} = V_{cap}$, such that during the second stimulation pulse the stimulation is stopped when the same amount of charge is reached. In this mode the user has better control over the first stimulation pulse.

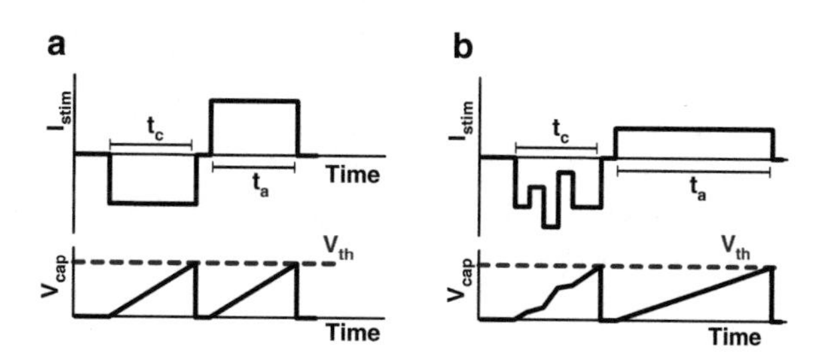

Fig. 6.2 Examples of the waveforms of I_{load} and V_{cap} from Fig. 6.1 during a stimulation cycle. In (**a**) a symmetrical constant current stimulation is shown, while in (**b**) an arbitrary asymmetrical stimulation waveform is depicted

Table 6.1 The system requirements of the arbitrary waveform stimulator that is designed in this chapter

Description	Value
Electrode impedance	$5\,\mathrm{k}\Omega < R_s < 20\,\mathrm{k}\Omega$
Amplitude range	$10\,\mu\mathrm{A} < I_{stim} < 1\,\mathrm{mA}$
Pulsewidth range	$100\,\mu\mathrm{s} < t_{stim} < 1\,\mathrm{ms}$
Pulse shape	Arbitrary (DAC with sampling time $t_s > 10\,\mu\mathrm{s}$)

The requirements are adopted from the application described in Sect. 6.4

In Fig. 6.2 the operating principle is illustrated by sketching the waveforms of I_{stim} and V_{cap}. In Fig. 6.2a a constant current stimulation is given and as can be seen the value of V_{cap} increases linearly until it reaches the value of V_{th}. In Fig. 6.2b the arbitrary waveform capabilities of the system are shown. Also, this example shows that it is possible to use asymmetrical biphasic stimulation schemes, while still achieving charge balanced operation.

The proposed topology has several advantages. Contrary to the implementation with a sense resistor, it is noted that the integrator can be single ended instead of differential. This relaxes the requirements for the implementation of the integrator. Furthermore, the current efficiency $N/(N+1)$ can be made high by selecting a high value for N.

The H-bridge based topology (S_1–S_4) was adopted to relax the requirements on mismatch. During both stimulation phases, I_{stim} is copied and integrated using the same circuitry. A constant mismatch in N or C does therefore not affect the *relative* mismatch of the charge between the two phases. It is important to realize that the absolute value of the charge does not need to be defined accurately, because this value is set empirically by the user.

A drawback of the proposed system is the limited capability for efficient multichannel operation. In order to implement multiple independent stimulation channels, the complete system of Fig. 6.1 needs to be duplicated.

The application for which this stimulator system is designed is discussed in detail in Sect. 6.4. The specifications for the stimulator are summarized in Table 6.1. The system is designed to connect to electrodes manufactured by Plastics One Inc (Roanoke, VA-USA). Electrode pairs that consist of two twisted insulated stainless steel wires with bared ends are used, which have a diameter of 0.01 inch. The series resistance of these electrodes is assumed to be $5\,\mathrm{k}\Omega < R_s < 20\,\mathrm{k}\Omega$.

The stimulation intensity required is $I_{stim} < 1\,\text{mA}$ and it is required to set the stimulation intensities with steps of $10\,\mu\text{A}$. The pulse width of a stimulation pulse is specified to be $t_{stim} < 1\,\text{ms}$, which sets the maximum charge contents of a stimulation pulse to $I_{stim}t_{stim} = 1\,\mu\text{C}$. The arbitrary waveform is assumed to be generated using a digital to analog converter (DAC) that uses a minimum sampling time of $t_s = 10\,\mu\text{s}$.

6.2 IC Circuit Design

The design discussed in this section has been simulated using the On Semiconductor (formerly AMI Semiconductor) I3T25 0.35 μm technology that includes options for high voltage DMOS transistors: besides the standard 3.3 V Low Voltage (LV) devices, the technology offers High Voltage (HV) DMOS devices that support up to 18 V drain voltages.

An overview of the system design is depicted in Fig. 6.3 in which two feedback loops can be identified. The design presented here has a voltage steered output and the voltage over the load is regulated by controlling the input of the driver block by means of the first feedback loop using the amplifier. The second feedback loop stops the stimulation as soon as the logic block has detected that the injected amount of charge has reached the threshold.

6.2.1 *Driver*

The driver is responsible for generating the stimulation current I_{stim} and the scaled copy $I_{int} = I_{stim}/N$ that is fed to the integrator. A straightforward implementation of this functionality is a standard cascoded current mirror as is given in Fig. 6.4a. The voltage V_g regulates I_{stim}, while the size ratio of the transistors ensures that I_{int} is N times smaller.

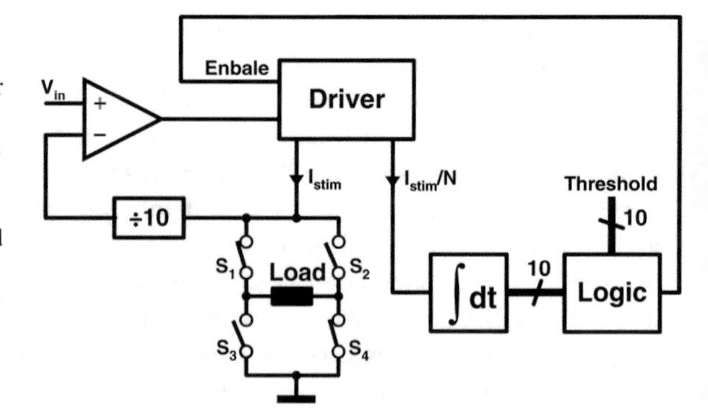

Fig. 6.3 System design of the voltage controlled arbitrary waveform stimulator system. The integrator converts the input current to a 10 bits digital signal by counting the number of periods in a current controlled oscillator. The digital value represents the charge injected during the stimulation pulse

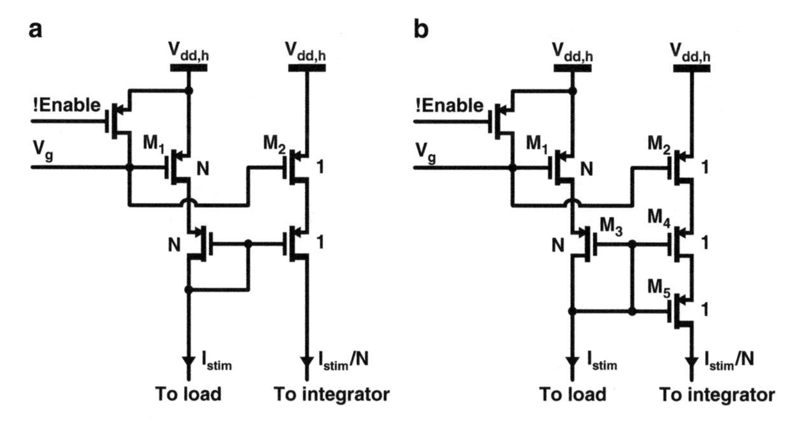

Fig. 6.4 Two implementations for the driver block. In (**a**) a simple cascoded current mirror is used to generate I_{stim} and I_{int}, while in (**b**) an implementation is depicted that minimizes the use of HV transistors

Fig. 6.5 Normalized current ratio I_{stim}/I_{int} as a function of the output voltage V_{out} over a load $R_{load} = 10\,\text{k}\Omega$, while the branch of I_{int} is connected to ground

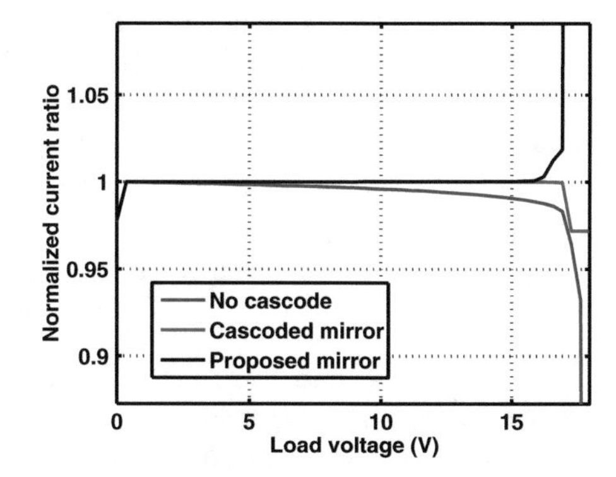

The cascode stage is used to compensate for the channel length modulation effect, which will change the ratio N of the output currents as a function of the load voltage. This effect is shown in Fig. 6.5. The voltage V_{out} over a $10\,\text{k}\Omega$ load was varied, while the integrator branch was connected to ground. Indeed the cascoded implementation helps to keep N constant for a longer time.

One of the drawbacks of the design of Fig. 6.4a is that it requires two high voltage (HV) transistors with size N. To reduce area consumption, the driver circuit in Fig. 6.4b converts the cascode to low voltage (LV) transistors. This is possible because the voltage drop over M_3 is always low (it is diode connected). One more additional HV transistor of size 1 is needed at the integrator output to prevent a high voltage drop over the LV transistor in this branch. Comparing Figs. 6.4a and b, it can be seen that the total number of HV transistors is reduced from $2N + 2$ towards $N + 2$, which can lead to a significant area reduction for high N.

The proposed solution comes at the price of a reduced voltage compliance of the driver: the voltage drop over the driver in the stimulation branch is now $V_{ds,1} +$

$V_{ds,3} = V_{ds,1} + V_{ds,4} + V_{gs,5}$. This is seen in Fig. 6.5 in which the ratio N becomes distorted from $V_{out} > 16$ V as opposed to $V_{out} > 16.9$ V for the cascoded case.

Since power efficiency is not a design goal of this particular system and since no adaptive supply mechanism is implemented, the compliance voltage is not a big concern. If power efficiency is to become important, an adaptive power supply can be used to adapt the supply voltage to the load [3]. In this case the compliance voltage should also be minimized, which can be achieved by using an active feedback loop, such as proposed in [4].

6.2.2 *Integrator Design*

The integrator design discussed in this section, as introduced in [5], converts the input current to the time domain, which allows for an implementation with low power operation and a high dynamic range.

The circuit implementation is shown in Fig. 6.6. The current I_{int} is converted to a periodic signal by means of a current controlled oscillator: each period at the output of the oscillator corresponds to a particular amount of charge. This periodic signal is subsequently fed into a counter to keep track of the number of injected charge packets. The dynamic range required at the output of the integrator is found by:

$$\mathrm{DR} = \frac{Q_{max}}{Q_{min}} = \frac{I_{stim,\max} t_{stim,\max}}{I_{stim,\min} t_{stim,\min}} \tag{6.1}$$

Using the values from Table 6.1 it is found that DR = 60 dB and hence a 10bit counter is used. The dynamic range of the integrator can be made arbitrarily large by increasing the number of bits of the counter.

The design proposed here is based on the threshold compensated inverter introduced in [6]. The basic idea is to have an inverter for which the threshold voltage can be set independent of V_{dd} and process variations. In this design the

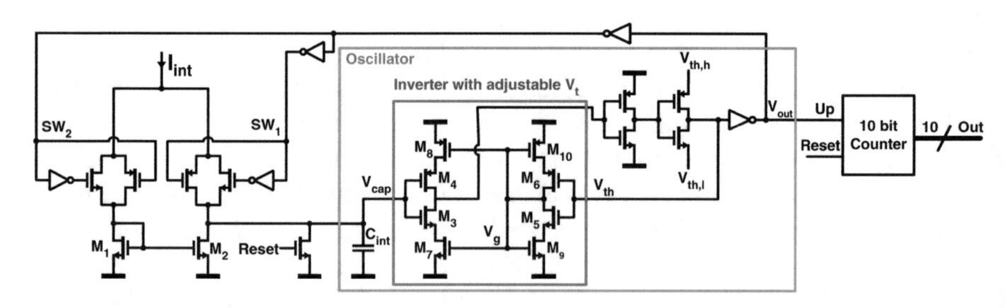

Fig. 6.6 Proposed schematic of the integrator circuit. The input current is converted to the time domain using a current controlled relaxation oscillator, which uses a Schmitt trigger that uses a threshold compensated inverter. The output is processed in the digital domain by counting the periods of the oscillator that correspond to a certain amount of charge

threshold is not only set, it is also varied between $V_{th,h}$ and $V_{th,l}$ to construct the functionality of a Schmitt trigger.

The oscillator integrates the input current I_{int} into a voltage V_{cap} across capacitor C. By enabling SW_1 and disabling SW_2, the voltage V_{cap} is increasing until it reaches threshold $V_{th,h}$. This will cause SW_1 and SW_2 to toggle, which will reverse the direction of the current through the capacitor by feeding it through the current mirror M_1–M_2. Now V_{cap} will decrease, until the low threshold $V_{th,l}$ is reached an SW_1 and SW_2 flip back again.

The circuit implementing the threshold compensated inverter is highlighted in Fig. 6.6. The basic inverter is formed by transistors M_3 and M_4. Transistors M_5 and M_6 are duplicates of M_3 and M_4 with the required V_{th} at their inputs. The output voltage of these transistors determines the gate voltage of transistors M_7–M_{10}. Transistors M_9 and M_{10} form a feedback loop with M_5 and M_6 to bias the V_{th} of M_3 and M_4 to the required value using M_7 and M_8. Note that the correct working principle of this circuit relies on the matching of transistor pairs M_3–M_5, M_4–M_6, M_7–M_9, and M_8–M_{10}.

Depending on the value of V_{th} there is a DC current flowing from V_{dd} to ground through M_{10}, M_6, M_5, and M_9, potentially leading to high static power consumption. Two possible solutions exist:

- Choose V_{th} to be close to ground or V_{dd}. In this case V_{th} will be far away from the 'natural' V_{th} of inverter M_3–M_4. This means that either M_{10} or M_9 will be in weak inversion, yielding a low current. For this particular application, a V_{th} which is close to V_{dd} or gnd is beneficial, because in this way $V_{swing} = V_{th,h} - V_{th,l}$ over the capacitor is maximized and therefore the full charge storing capability of the capacitor is used.
- The length of transistors M_7, M_8, M_9, and M_{10} can be increased, yielding a lower static current through the right branch of the circuit.

These methods can be used in combination with each other. However, they will have some negative consequences on the performance of the circuit. By increasing the length of transistors M_7 and M_8, the speed of the circuit is decreased: it will take longer to switch the output ((dis)charging the next inverter). Increasing the length also increases the capacitive load at node V_g.

It turns out it is possible to make M_7 and M_8 much shorter than M_9 and M_{10} while still achieving sufficient accuracy of the threshold voltages. Consider the case when V_{th} is chosen close to 0 V. In this case M_7 and M_9 will be in saturation, while M_8 and M_{10} will be in triode. For a saturated transistor the following equation holds:

$$I_d = \mu C_{ox} \frac{W}{L} \left(V_{gs} - V_t\right)^2 (1 + \lambda V_{ds}) \tag{6.2}$$

Here μ represents the effective mobility, C_{ox} the gate oxide capacitance per unit area, W the width, L the length, V_t the threshold voltage, and λ is the channel length modulation parameter of the transistor. If it is assumed that $\lambda = 0$, the current I_d is decreased proportionally for an increase in L, irrespective of V_{ds}. For the transistors in triode it holds, assuming $V_{ds} \ll 2(V_{gs} - V_t)$:

$$V_{ds} = \frac{L}{\mu C_{ox} W (V_{gs} - V_t)} I_d \tag{6.3}$$

This shows that although the L is increased, the saturated transistor makes the I_d decrease with the same factor, yielding the same V_{ds} for the triode transistor. This means V_{th} should still be the same while M_9 and M_{10} can be much larger than M_7 and M_8. A similar reasoning can be made for the situation in which V_{th} is chosen close to V_{dd}.

In Fig. 6.7a the DC response is given for several values of V_{th}. For the solid lines transistors M_7, M_8, M_9, and M_{10} all have $L = 20\,\mu\text{m}$, while for the dashed lines M_7 and M_8 were given $L = 1\,\mu\text{m}$. As can be seen, the threshold voltages of the symmetrically sized circuit indeed correspond to the values set. For the asymmetrical circuit some obvious deviations are clear. However, the values relatively close to V_{dd} and *gnd* show only minor (<20 %) deviations. It was chosen to take $V_{th,l} = 0.5\,\text{V}$ and $V_{th,h} = V_{dd} - 0.5 = 2.8\,\text{V}$, which makes $V_{swing} = 2(2.8 - 0.5) = 4.6\,\text{V}$.

In Fig. 6.7b the simulation results of the complete integrator are presented for three different input currents that cover the complete input current range. Both the triangular voltage V_{cap} and V_{out} are depicted and as can be seen the frequency of these signals depends on the input current I_{int}. The frequency is not completely linear with the input current, due to the delay introduced by the switching of SW_1 and SW_2. This deviation will lead to small charge imbalance when there is a large difference in amplitude between the two stimulation phases. A larger value for C would reduce this error.

The static power consumption is mainly dominated by the static current through the M_5–M_6–M_9–M_{10}-branch of the threshold compensated inverter. When the integrator is reset ($V_{cap} = 0\,\text{V}$) the power consumption is simulated to be 171 nW.

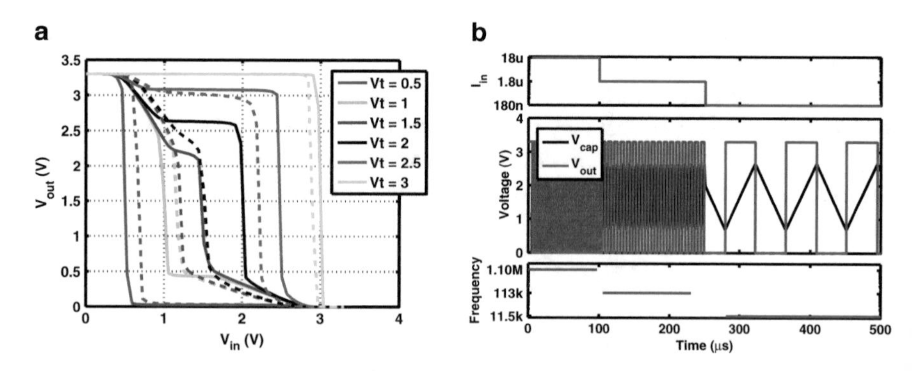

Fig. 6.7 Simulation results from the integrator system. In (**a**) the DC response of the V_t compensated inverter for a symmetric (*solid lines*) and asymmetric (*dashed lines*) topology is shown. It is seen that for $V_{th} \approx gnd$ and $V_{th} \approx V_{dd}$ the response of both systems is almost identical. In (**b**) the simulations of the complete integrator are shown

When the integrator is active, the power consumption increases, mainly due to current through the M_3–M_4–M_7–M_8-branch of the threshold compensated inverter. The current depends on the threshold voltages set at the Schmitt trigger, since this determines the level of inversion of M_5 and M_6 by V_g. For 0.5 and 2.5 V the current consumption increases to 800 nA and 4.1 μA, respectively. This can be decreased by shifting the threshold voltages more to 0 V and V_{dd} or by increasing the length of M_5 and M_6 at the cost of speed since it will make the output current smaller.

6.2.3 *Amplifier*

The amplifier is implemented using the circuit in Fig. 6.8. At the input stage a differential PMOS stage (M_1–M_2) is required to accommodate the values of $0\,\text{V} < V_{fb} < 1.8\,\text{V}$ due to the 10 times attenuation of the feedback network R-$9R$. These resistors are implemented using high ohmic poly resistors with $R \approx 50\,\text{k}\Omega$.

The noise and offset introduced in the amplifier are of little concern for this system: the voltage waveform over the load does not need to be very well defined as was outlined in Chap. 5. Small transistors ($W/L = 4/4\,\mu\text{m}$) were chosen for M_1 and M_2 and the bias $I_{bias1} = 0.3\,\mu\text{A}$ was found to be sufficient to drive the load of the second stage.

The second stage, transistor M_5, needs to drive the gate of the driver, which requires an HV signal. Therefore an HV transistor with HV biasing is required for which the values are determined by the required drive capability of the load. To switch the output stage from $I_{stim} = 0$ to $I_{stim} = 1$ mA, it was found that a charge of 35 pC needs to be delivered by the amplifier to the gates of the drivers. To be able to make this worst case transition within the minimum sampling interval of V_{in} of $t_s = 10\,\mu\text{s}$, a bias current of $I_{bias2} = 4\,\mu\text{A}$ was chosen.

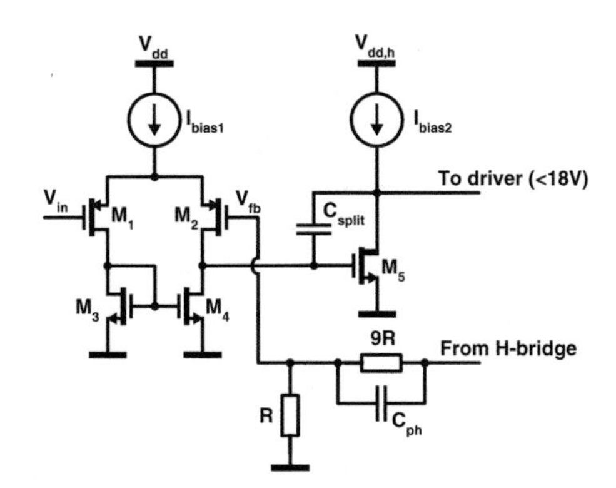

Fig. 6.8 Circuit implementation of the amplifier used for the voltage control of the tissue. The second amplifier stage M_5 is needed to shift the signal to the HV domain

6.2.4 Full System Simulations

Some examples of simulation results of the complete system are depicted in Fig. 6.9. For all these situations the load was chosen to be $R_{load} = 10\,\text{k}\Omega$ and $C_{load} = 500\,\text{nF}$. The charge threshold was selected to be equal to 255 charge packets, which is equivalent to about 220 nC.

In Fig. 6.9a a constant stimulation voltage of 3.5 V is used. As can be seen the stimulation current is exponentially decreasing during both phases. The charge on C_{load} is first increased to the expected 220 nC after which it is brought back to a value close to zero.

In Fig. 6.9b a burst stimulation pattern is used in which stimulation is switched on and off with a high frequency. Here pulses of 50 μs were chosen with a duty cycle of 50 %. In Fig. 6.9c a random signal is played by using a white noise signal that is generated by linear interpolation of random samples with a normal distribution that

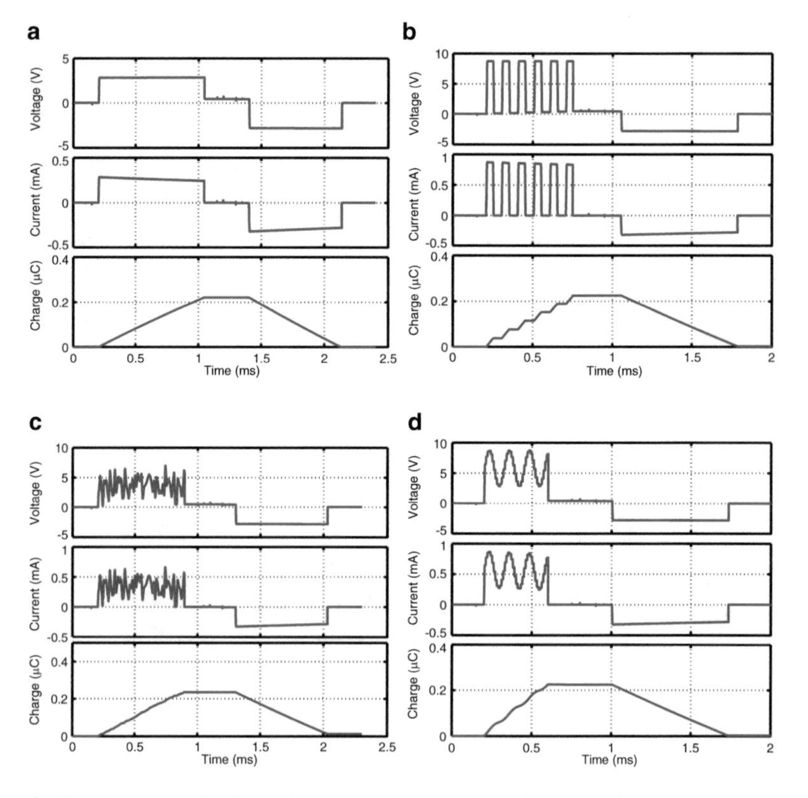

Fig. 6.9 Simulation results illustrating the arbitrary stimulation waveforms produced by the system. All waveforms are biphasic and charge balanced, use a load of $R_{load} = 10\,\text{k}\Omega$ and $C_{load} = 500\,\text{nF}$. The following waveforms are used for the first stimulation phase: (**a**) symmetrical constant voltage ($V_{tis} = 3.5\,\text{V}$), (**b**) burst stimulation, (**c**) stochastic stimulation, and (**d**) a sinusoidal signal. The second stimulation phase is always a constant voltage of 3.5 V

are generated every 10 μs. In Fig. 6.9d the situation is simulated in which a DAC would generate the stimulation signal: a sampled 8 kHz sinusoid is generated with a sampling speed of 100 kHz. In the last three cases the second stimulation phase was chosen to have a constant voltage of 3.5 V. This makes all these stimulation waveforms asymmetrical.

The charge mismatch that is found with the stimulation waveforms in Fig. 6.9 depends on the waveform used. The simulated mismatch is found to be 32 pC (0.145 %), 2.48 nC (1.12 %), 12.94 nC (5.8 %), and 2.04 nC (0.93 %) for Figs. 6.9a–d, respectively. These charge mismatch percentages are relatively high in comparison with other stimulator systems such as [7], which reports a mismatch as low as 6 ppm. However, these systems only support symmetrical constant current stimulation and rely on current matching circuit techniques. These conditions are not valid for arbitrary waveform stimulators and therefore it is hard to compare these systems.

The mismatch percentages indicate that it would be necessary to include additional charge balancing techniques to make the system safe for clinical applications. Either passive discharging can be used if the stimulation rate is low enough or active charge balancing schemes such as pulse insertion are possible.

6.3 Discrete Realization

As explained in the introduction of this chapter, it was decided to make a discrete component realization of the arbitrary waveform stimulator as opposed to an IC realization, to allow for rapid prototyping. Furthermore, power consumption and size were not important design requirements for the percutaneous application.

The discrete realization uses a similar system structure as the IC realization, but differs at a few points:

- It uses a current steered output instead of a voltage steered output. This causes the feedback loop that controls the stimulation waveform to change. A current control loop based on the principle outlined in [4] was used.

  In [4] a double loop feedback topology was proposed to accurately control the output current of a stimulator, while only a single transistor is needed at the output branch. The advantage of this structure is that a single transistor can minimize the compliance voltage of the output stage, which is potentially beneficial to maximize power efficiency if the circuit is combined with an adaptive supply configuration. For this implementation the compliance voltage is not very important and hence it was decided to replace one of the feedback loops using a simple cascode stage.
- The integrator is implemented in a simpler fashion. Because high valued capacitors are readily available for discrete realizations, it is not necessary to convert the signal to the time domain. The 60 dB dynamic range was accommodated by scaling the value of the integrating capacitor C by using a capacitor bank.

- The system needs to be able to drive 8 electrodes. From these electrodes any arbitrary number should be set to be the cathode, while another arbitrary number will be the anode.

  The system will only implement a single stimulation channel for which the total amount of charge is balanced. If the impedance of the electrodes changes with respect to each other during a stimulation pulse, it is possible that there is charge imbalance at each individual electrode. Charge balancing is guaranteed however if only one anode and one cathode are used.

In the following two sections the circuit implementation and the measurement results of the discrete version are presented.

6.3.1 Circuit Design

A simplified circuit implementation of the discrete arbitrary waveform stimulator is depicted in Fig. 6.10. The stimulation current is regulated by means of a Digital to Analog Converter (DAC1) that generates a voltage V_{in}. The input current I_{in} is created by keeping the voltage over resistor R_1 equal to V_{in} by means of the feedback loop around M_1 and O_1. This input current is copied to I_{stim} by means of a current feedback loop: at Node N_1 the error current I_e is regulated towards zero by adjusting the base voltage of transistors Q_1. To minimize the early effect at Q_1, the current mirror is cascoded using M_2. Transistors Q_1 have a 1:1 size ratio and hence for this discrete realization $N = 1$. Transistor M_3 is used to enable/disable the stimulation: when it is switched on, the gate voltage of M_1 is forced to 0 V, which will make $I_{in} = 0$ A.

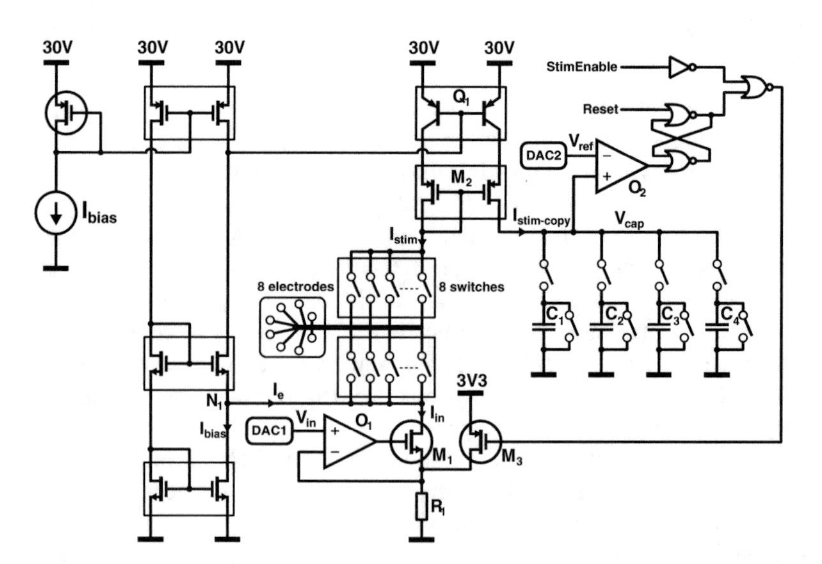

Fig. 6.10 Simplified circuit diagram of the discrete implementation of the arbitrary waveform stimulator system

The stimulation current I_{stim} is fed through any combination of the 8 electrodes set by the switch arrays. Each switch array is implemented using an octal SPST (Single Pole Single Throw) switch array IC (MAX335) that uses an SPI interface. The biphasic stimulation scheme is implemented using the H-bridge topology: after the first stimulation cycle is finished, the switch configuration of both arrays is reversed, which also reverses the stimulation current direction through the electrodes.

The current I_{int} is fed to the integrator that is implemented using a combination of capacitors C_1–C_4 to obtain the charge: $V_{cap} = C^{-1} \int I_{int} dt$. Using the switches any combination of capacitors C_1–C_4 can be selected in order to maximize V_{cap} based on the approximate charge contents of a stimulation pulse. Comparator O_2 can be used to disable the stimulation using the digital logic upon reaching a certain charge threshold that is set by V_{ref}, generated by DAC2. The signals 'StimEnable' and 'Reset' are used to enable the stimulation and to reset the latch after the charge threshold is reached, respectively.

A typical stimulation cycle will now work as follows. First the switch arrays are put in position for the first stimulation phase. To fully inject the first stimulation waveform, O_2 is prevented from stopping the stimulation by setting the value of V_{ref} to V_{dd}. Subsequently the stimulation is enabled by the 'StimEnable' signal, while DAC1 is generating the stimulation waveform. Upon finishing, the value of V_{cap} is sampled by an Analog to Digital Converter (ADC) and is copied on DAC2.

Subsequently, the integrator is reset and the configuration of the switch arrays is reversed to prepare for the second stimulation phase. Upon enabling using 'StimEnable' and setting the desired waveform on DAC1, the stimulation will automatically stop when V_{cap} reaches V_{ref} by means of O_2, which will mean that the charge has been balanced.

The circuit presented in Fig. 6.10 is controlled by the Beaglebone credit-card sized development board that includes the AM335 × 720 MHz ARM Cortex-A8 microprocessor. The 3.3 V supply from this board is used as the LV power supply, while it generates an HV 30 V power supply is created using a boost converter. The internal ADC of the processor is used to sample V_{cap}, while DAC1 and DAC2 are implemented using the LTC2602 IC, which uses an SPI interface.

6.3.2 Measurement Results

The circuit from Fig. 6.10, along with support circuitry such as voltage regulators and level converters, is realized on PCB and connected to the Beaglebone development board, as shown in Fig. 6.11. A few additional components were added to the circuit to increase the safety in a practical situation, such as 2.2 μF coupling capacitors and Schottky diodes to ensure the current to flow in the correct direction through the load.

A series *RC* load ($R = 8.2\,\mathrm{k}\Omega$, $C = 220\,\mathrm{nF}$) was connected between two electrodes and a variety of asymmetrical stimulation waveforms was programmed in the Beaglebone. A single stimulation waveform was passed through the initially

Fig. 6.11 Discrete realization of the arbitrary waveform stimulator system. The PCB at the bottom is the Beaglebone microprocessor platform that is used to control the upper PCB. This PCB includes the stimulator system as discussed in this chapter, along with the necessary support circuitry

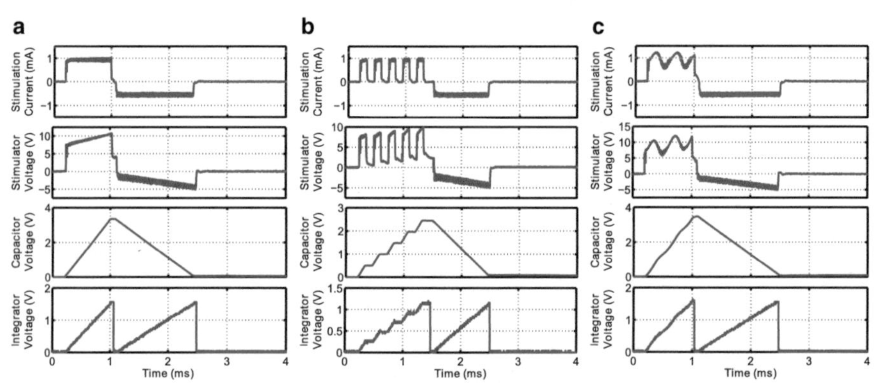

Fig. 6.12 Measurement results of the discrete realization of the arbitrary waveform generator as described in Fig. 6.10. All subfigures show a biphasic asymmetric stimulation scheme with (**a**) a constant current, (**b**) a burst waveform, and (**c**) a sinusoidal waveform

uncharged load and in Fig. 6.12 the measurement results are depicted. The currents through and voltages across the load were captured by using a Hewlett Packard 1142A differential probe. Mismatch is 12 nC (1.5 %), 19 nC (3.5 %), and 12 nC (1.6 %) for the constant, burst, and sinusoidal case, respectively. These percentages are comparable to the results from the IC simulation in Fig. 6.9.

The measured charge mismatch is acceptable for the application that is discussed in Sect. 6.4. For this application there will be a limited amount of stimulation cycles, with a large time (in the order of seconds) between them. This will give enough time to discharge the load capacitor using passive discharging. Therefore it is not needed to include active charge balancing schemes.

6.4 Application: Multimodal Stimulation for the Reduction of Tinnitus

As described in [8], tinnitus is a condition in which "a subject experiences the perception of sounds in the absence of a corresponding external acoustic stimulus." Although the underlying causes are not well understood and are found to be multifactoral, in many cases tinnitus is believed to be linked to pathological changes in the auditory pathway, such as hearing loss. Apparent correlations between the hearing loss of the subject and the perceived tinnitus spectrum have been reported [9]. One of the causes of tinnitus is therefore hypothesized to be a reaction of the nervous system due to the change in the auditory pathway, comparable to phantom pain perceptions after limb amputation.

Many treatment methods exist such as psychological treatment, auditory stimulation, pharmacological treatments, and brain stimulation [8]. Most treatment methods use only one of these aspects simultaneously. There are, for example, many studies on using audio treatment for tinnitus treatment. Encouraging results were obtained in animal experiments [10], but human trials show a much lower efficacy [11].

The central idea behind the proposed treatment methodology is to combine multiple modalities into one treatment. In this case auditory stimulation is combined with electrical stimulation. The electrical stimulation is used to enhance the effect of the auditory stimulation by tapping in on the positive and negative reward mechanisms of the nervous system. The reward system, which is involved in learning via Pavlovian and operant conditioning includes the Ventral Striatum, the Nucleus Accumbens, and the Habenula. The Nucleus Accumbens is considered to be involved in the positive reward system of the human nervous system [12]. On the contrary, the Habenula is thought to be involved in providing the nervous system with a negative reward and stimulation of the Habenula can therefore induce a negative feedback [13].

The proposed stimulation strategy is the following. If the auditory stimulus has a spectrum that falls outside the tinnitus band, a positive stimulus is given by stimulating the Nucleus Accumbens. In this way the nervous system is trained to consider the non-tinnitus frequencies as something positive. At the same time a negative feedback is induced by stimulating the Habenula when the auditory stimulus falls within the tinnitus spectrum. This will train the nervous system to consider the tinnitus frequencies as something negative. It is hypothesized that this will eventually decrease or eliminate the unwanted response that leads to tinnitus.

This technique can also be used for other pathologies, such as addictions. By giving a negative reward to the nervous system while presenting the subject with an input that is linked with the addiction, the brain can be reconditioned to consider the addiction as something bad. For example, an alcohol addict can be provided with images, smells, or even real alcoholic drinks, while a negative reward is presented. Similarly, a positive reward can be presented while providing the subject with non-alcoholic beverages.

6.4.1 Materials

The electrodes used are Plastics One MS303/2-B/SPC 2 channel twisted wire electrodes. Each stainless steel wire (diameter 0.2 mm) is coated with polyimide, except for the tips (0.2 mm) where the contact with the tissue is made. To accommodate

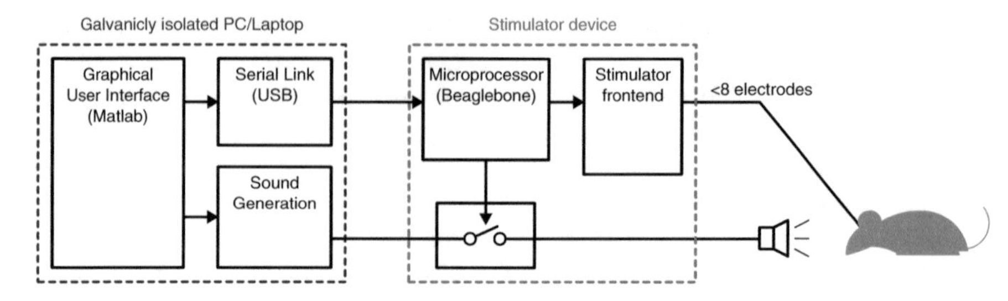

Fig. 6.13 Overview of the system topology used for the experiments: a computer is using Matlab to generate the sound stimulus and to provide the user with a GUI. The computer subsequently connects to the stimulation device which synchronizes the electrical stimulation with the audio signal, which are both delivered to the subject

future setups, it was decided to equip the stimulator with 8 electrode contacts, from which an arbitrary number can be selected to be the anode and another arbitrary number to be the cathode (as discussed in Sect. 6.3).

The stimulation settings were already outlined in Table 6.1. On top of that, the stimulator needs to be able to deliver burst stimulation [14]: this means that a number of stimulation cycles (typically 5) are repeated shortly after each other. Two different burst modes exist: either a complete stimulation cycle (including the charge balancing phase) is repeated or only the first stimulation cycle is repeated (like the waveform from Fig. 6.12b). In the first realization only the first option is implemented, but thanks to the arbitrary waveform capabilities, the second option can be implemented relatively easily in future realizations.

The audio presentation should be either single tone or noise-like. When single tone mode is used, the electrical stimulation is adjusted based on whether the frequency of the tone falls in- or outside the tinnitus range. When noise is used, the frequency contents of the noise are filtered based on the frequency of the tinnitus.

The complete system that is able to meet the electrical and auditory requirements discussed above is depicted in Fig. 6.13. The arbitrary waveform stimulation as discussed in Sect. 6.3 is used to construct a neural stimulator device that is able to synchronize, i.e. pair an auditory stimulus with the electrical stimulation. The system is controlled using a PC or laptop that is galvanically isolated from the ground for safety reasons.

The computer sends the stimulation settings to the stimulator device using a serial connection. When a stimulation cycle is initiated, the PC will first generate the auditory stimulus. The audio signal is synchronized with the electrical stimulation by the microprocessor that will close the switch (ADG621) simultaneously with the start of the electrical stimulation.

Using a graphical user interface (GUI) as depicted in Fig. 6.14, the user can adjust all the necessary parameters for stimulation. In this realization two stimulation patterns can be generated. Pattern 1 combines auditory stimulation with electrical stimulation and is used to stimulate the Nucleus Accumbens while a non-tinnitus frequency is presented to the subject. Pattern 2 uses auditory stimulation exclusively

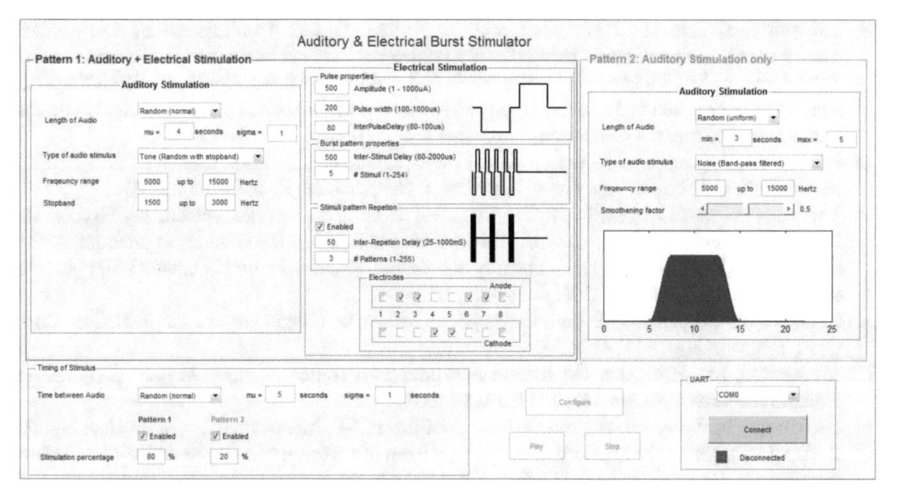

Fig. 6.14 Graphical user interface used to control the synchronized audio/electrical stimulation system

and can be used to present tinnitus frequencies to the subject. In the future the capabilities of this GUI can be extended to include another independent electrical stimulation for Pattern 2.

References

1. Xiang, F., Wills, J., Granacki, J., LaCoss, J., Arakelian, A., Weiland, J.: Novel charge-metering stimulus amplifier for biomimetic implantable prosthesis. IEEE International Symposium on Circuits and Systems, pp. 569–572 (2007)
2. van Dongen, M.N., Serdijn, W.A.: Design of a versatile voltage based output stage for implantable neural stimulators. IEEE First Latin American Symposium on Circuits and Systems (2010)
3. Noorsal, E., Sooksood, K., Xu, H., Hornig, R., Becker, J., Ortmanns, M.: A neural stimulator frontend with high-voltage compliance and programmable pulse shape for epiretinal implants. IEEE J. Solid-State Circuits **47**(1), 244–256 (2012)
4. Sawigun, C., Ngamkham, W., van Dongen, M.N., Serdijn, W.A.: A least-voltage drop high output resistance current source for neural stimulation. IEEE Biomedical Circuits and Systems Conference (BioCAS), pp. 110–113 (2010)
5. van Dongen, M.N., Serdijn, W.A.: Design of a low power 100 dB dynamic range integrator for an implantable neural stimulator. IEEE Biomedical Circuits and Systems Conference (BioCAS), pp. 158–161 (2010)
6. Tan, M.T., Chang, J.S., Tong, Y.C.: A process-independent threshold voltage inverter-comparator for pulse width modulation applications. Proceedings of IEEE International Conference on Electronics, Circuits and Systems, vol. 3, pp. 1201–1204 (1999)
7. Site, J.J., Sarpeshkar, R.: A low-power blocking-capacitor-free charge-balanced electrode-stimulator chip with less than 6 nA DC error for 1-mA full-scale stimulation. IEEE Trans. Biomed. Circuits Syst. **1**(3), 172–183 (2007)

8. Langguth, B., Kreuzer, P.M., Kleinjung, T., Ridder, D. De: Tinnitus: causes and clinical management. Lancet Neurol. **12**(9), 920–930 (2013)
9. Schecklmann, M., Vielsmeier, V., Steffens, T., Landgrebe, M., Langguth, B., Kleinjung, T.: Relationship between audiometric slope and tinnitus pitch in tinnitus patients: insights into the mechanisms of tinnitus generation. PLoS One **7**(4) (2012)
10. Norea, A.J., Eggermont, J.J.: Enriched acoustic environment after noise trauma reduces hearing loss and prevents cortical map reorganization. J. Neurosci. **25**(3), 699–705 (2005)
11. Vanneste, S., van Dongen, M.N., De Vree, B., Hiseni, S., van der Velden, E., Strydis, C., Joos, K., Norena, A., Serdijn, W.A., De Ridder, D.: Does enriched acoustic environment in humans abolish chronic tinnitus clinically and electrophysiologically? A double blind placebo controlled study. Hear. Res. **296**, 141–148 (2013)
12. Knutson, B., Cooper, J.C.: Functional magnetic resonance imaging of reward prediction. Curr. Opin. Neurol. **18**(4), 411–417 (2005)
13. Matsumoto, M., Hikosaka, O.: Lateral habenula as a source of negative reward signals in dopamine neurons. Nature **447**, 1111–1115 (2007)
14. De Ridder, D., Vanneste, S., Loo, E., van der Plazier, M., Menovsky, T., van de Heyning, P.: Burst stimulation of the auditory cortex: a new form of neurostimulation for noise-like tinnitus suppression. Br. J. Neurosurg. **112**(6), 1289–1294 (2010)

Chapter 7
Switched-Mode High Frequency Stimulator Design

Abstract This chapter presents a neural stimulator system that employs a fundamentally different way of stimulating neural tissue compared to classical constant current stimulation. It uses the concept of switched-mode duty cycled stimulation as introduced in Chap. 4: a stimulation pulse is composed of a sequence of current pulses injected at a frequency of 1 MHz for which the duty cycle is used to control the stimulation intensity.

Implantable neural stimulators impose strict requirements on the power consumption, safety, and size of the system. The number of external components should be kept to a minimum to limit the size and increase the device safety, while the power efficiency should be maximized in order to limit the battery size.

Furthermore there is a trend towards an increasing number of stimulation channels. Some applications, such as cochlear implants [1] or retinal implants [2–4], need a high number of channels to accommodate a large amount of stimulation sites. Other applications, such as deep brain stimulation (DBS) or peripheral nerve stimulation (PNS) [5], use multiple channels to implement current steering [6, 7] to achieve more localized neuronal recruitment with fewer side effects.

Current source based stimulation is usually preferred in order to accurately control the amount of charge during a stimulation cycle. A high-level system architecture of a typical current based stimulator is shown in Fig. 7.1a: a power efficient switched voltage converter, here referred to as a dynamic supply, is used to control V_{dd} to supply the current source that generates the stimulation current I_{stim} [8, 9]. As can be seen this system uses at least two external components (the inductor L and the output capacitor C). Also, as will be shown, the power efficiency degrades when the system is operated in multichannel mode.

In this chapter an implementation is discussed that uses the dynamic power supply to stimulate the tissue directly [10], as shown in Fig. 7.1b. The output capacitor is omitted and hence only one external component is required. The exclusion of this capacitor leads to a fundamentally different stimulation principle: L is repeatedly discharged through the load. The stimulation waveform through the load consists of a train of high frequency current pulses, each of which contains a well defined amount of charge.

M. van Dongen, W. Serdijn, *Design of Efficient and Safe Neural Stimulators*,
Analog Circuits and Signal Processing, DOI 10.1007/978-3-319-28131-5_7

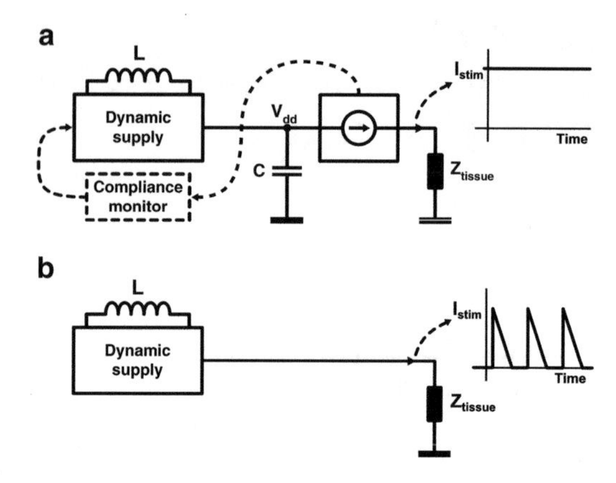

Fig. 7.1 High-level system architectures of (**a**) a typical (constant) current based stimulator system with adaptive supply and (**b**) the high frequency dynamic stimulator proposed in this work

The proposed system is particularly suitable to operate in multichannel mode. The inductor can be discharged in an alternated fashion through different electrodes with tailored stimulation intensities, making simultaneous and independent stimulation possible over multiple electrodes, without the need for additional external components. The advantage of the proposed system is that the power efficiency is almost not degraded when it is operated in multichannel mode, as opposed to state-of-the-art current based stimulators.

The chapter is organized as follows. In Sect. 7.1 the power efficiency of classical adaptive supply constant current stimulation is analyzed and it is shown how high frequency dynamic stimulation can improve the efficiency. In Sect. 7.2 the system design is discussed, emphasizing the digital control that enables the independent multichannel operation. Section 7.3 subsequently discusses the circuit design of some of the system blocks in detail. Finally in Sect. 7.4 the measurement results of a prototype IC realization are presented, comparing the power efficiency of the proposed system with state-of-the-art current source based systems.

7.1 High Frequency Dynamic Stimulation

7.1.1 Power Efficiency of Current Source Based Stimulators

The power efficiency of current source based stimulators is in general limited due to the voltage drop over the current driver. A popular way to increase the power efficiency is to adapt the power supply to the load voltage using a compliance monitor [2, 11]. A generic biphasic constant current stimulator setup is depicted in Fig. 7.2a with the load being modeled as a resistance R_{load} and capacitance C_{load} [12]. During stimulation there is a constant voltage drop $V_R = I_{stim}R_{load}$ over R_{load}, while the capacitor is charging towards $V_C = I_{stim}t_{stim}/C_{load}$.

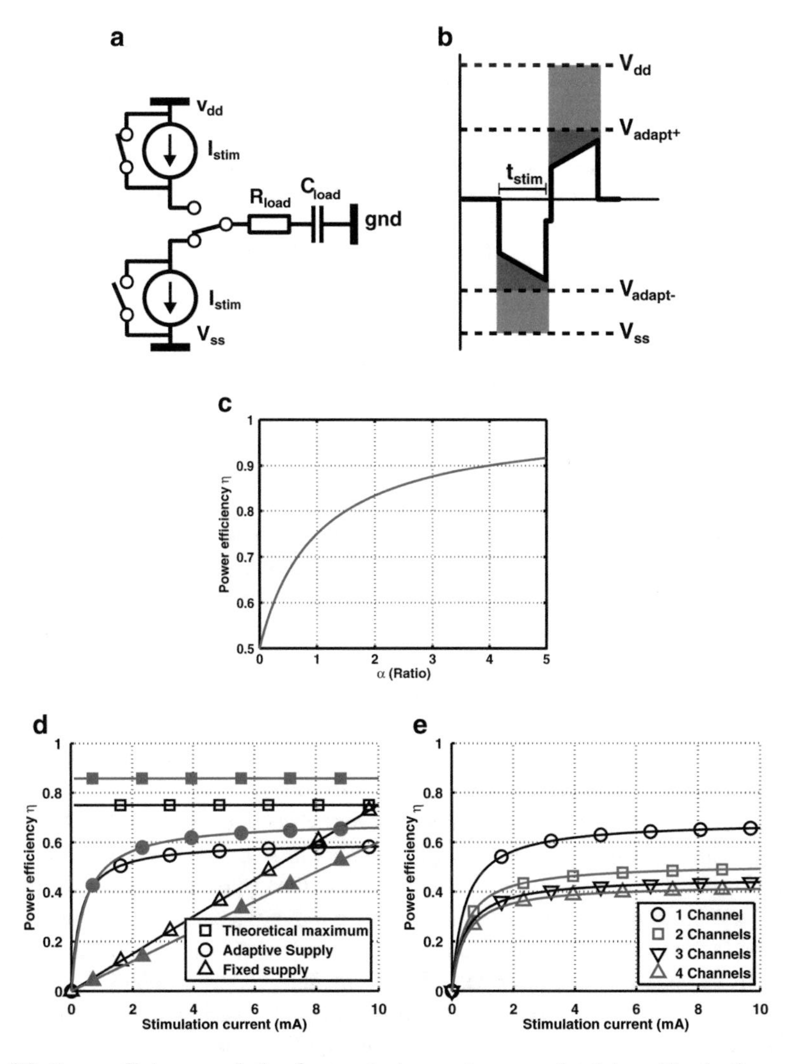

Fig. 7.2 Power efficiency analysis of a constant current source stimulator with adaptive power supply. In (**a**) a generic biphasic stimulator is depicted, in (**b**) the losses due to an adaptive supply stimulator (*dark gray*) and a fixed supply stimulator (*light gray*) are visualized. In (**c**) the theoretical maximum power efficiency is depicted as a function of $\alpha = V_R/V_C$. In (**d**) and (**e**) the power efficiency is depicted for a more realistic system. In (**d**) the *black lines* correspond to a load of $R_{load} = 500\,\Omega$, $C_{load} = 1\,\mu\text{F}$ with $t_{stim} = 200\,\mu\text{A}$ ($\alpha = 1$), while the *red lines* correspond to $R_{load} = 500\,\Omega$, $C_{load} = 10\,\mu\text{F}$ ($\alpha = 2.5$). It is seen in (**e**) that the efficiency drops dramatically when the system is operated with multiple channels

The efficiency of an adaptive supply system depends on the ratio $\alpha = V_R/V_C$. Referring to Fig. 7.2b, the load dissipates $E_{load} = I_{stim}V_R t_{stim} + 0.5V_C I_{stim}t_{stim}$. Assuming the supply will adapt to $V_{adapt} = V_R + V_C + V_{compl}$ in which V_{compl} is the minimum required compliance voltage for the current driver, the energy delivered by the source is $E_{supply} = \eta_{sup}^{-1} I_{stim} V_{adapt} t_{stim}$, in which η_{sup} is the efficiency of the adaptive supply generator. The power efficiency η_{adapt} is now found to be:

$$\eta_{adapt} = \frac{E_{load}}{E_{supply}} = \frac{\eta_{sup} V_C(0.5 + \alpha)}{V_C(1 + \alpha) + V_{compl}} \tag{7.1}$$

The theoretical maximum efficiency is found with an ideal adaptive supply generator $\eta_{sup} = 1$ and a current source with $V_{compl} = 0$:

$$\eta_{adapt,ideal} = \frac{0.5 + \alpha}{1 + \alpha} \tag{7.2}$$

This equation is plotted in Fig. 7.2c. Realistic values for η_{adapt} will be lower due to practical values of V_{compl} and η_{sup}. As an example a system is considered with $V_{compl} = 300\,\text{mV}$ [2] and $\eta_{sup} = 80\,\%$ [11], which is connected to a load of $R_{load} = 500\,\Omega$, $C_{load} = 1\,\mu\text{F}$, and $t_{stim} = 200\,\mu\text{A}$ ($\alpha = 1$). The efficiency as a function of I_{stim} is depicted in Fig. 7.2d by the black solid line. As can be seen the performance degrades as compared to the theoretical maximum, especially for low stimulation intensities. In the same Figure the efficiency can be compared with a load consisting of $R_{load} = 500\,\Omega$, $C_{load} = 10\,\mu\text{F}$ ($\alpha = 2.5$, red lines) and with a classical non-adaptive supply system with $V_{dd} = 10\,\text{V}$ (dashed lines).

The efficiency is even more degraded when the system is operated with multiple channels. Since there is only one supply voltage, this voltage needs to adapt to the channel with the highest $V_R + V_C$. This means that when other channels have a lower $V_R + V_C$, the efficiency is reduced. Such a reduction can be due to two factors. The first factor is impedance variation [13]: clinical values of the impedance spread of electrode contacts within individual DBS patients report standard deviations as high as $500\,\Omega$ (mean $1200\,\Omega$) [14]. The second factor is variation in the stimulation intensity, which is common in current steering applications.

As an example the effect of impedance variation on the efficiency in multichannel operation is considered. Multiple loads are stimulated simultaneously with $t_{stim} = 200\,\mu\text{A}$. Channel 1 has $R_{load,1} = 500\,\Omega$ and $C_{dl,1} = 10\,\mu\text{F}$, while Channel 2 has $R_{load,2} = 200\,\Omega$ and $C_{dl,2} = 10\,\mu\text{F}$. As can be seen in Fig. 7.2e, the power efficiency for this dual channel operation mode drops from 65 % down to 50 %, due to the reduced efficiency for Channel 2. Including even more channels with $R_{load,n} = 200\,\Omega$ will decrease the efficiency further to about 40 % for 4 simultaneous channels. A similar effect can be found for variations in stimulation current.

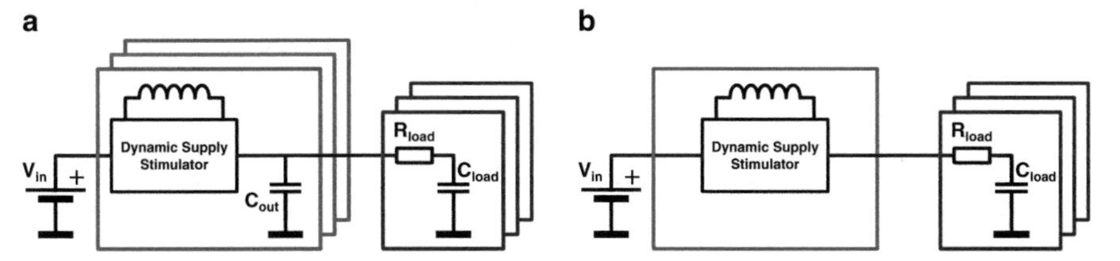

Fig. 7.3 System setup of dynamic power supply based stimulators. In (**a**) the system architecture is depicted of [15], while in (**b**) the proposed architecture is shown. Removal of the output capacitor C_{out} and using only a single dynamic supply allows for efficient multichannel stimulation

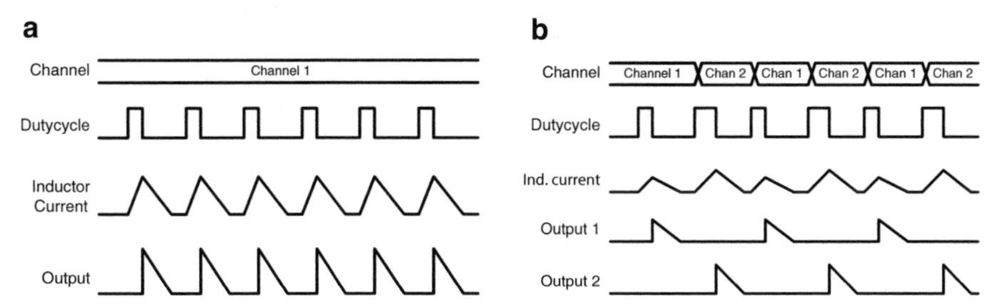

Fig. 7.4 Working principle of the proposed high frequency stimulator system. In (**a**) single channel operation is depicted, while in (**b**) simultaneous dual channel operation is shown in which the stimulus pulses are sent to the load in an alternating fashion

7.1.2 High Frequency Dynamic Stimulation

A system that is much less sensitive to α was introduced in [15] by using a dynamic supply to drive the load directly, as shown in Fig. 7.3a. The output voltage of the dynamic power supply is connected directly to the load, which means that the power efficiency is not dependent on α, but instead on the efficiency of the switched supply.

The system uses at least two external components: the inductor L for the dynamic power supply and a capacitor C_{out} to filter the switched output signal. This system does not scale well if multiple channels need to be controlled independently and simultaneously. Due to the filtering properties of C_{out} the voltage cannot be controlled for multiple channels individually. Furthermore, the stimulation is voltage steered, which means that charge is not controlled directly.

As was outlined in [16], it is proposed to remove C_{out} from the system as shown in Fig. 7.3b. A duty cycled signal is used to charge and subsequently discharge the inductor through the load as sketched in Fig. 7.4a. In Chap. 4 it was shown that this high frequency pulsating stimulation pattern is able to induce effective neuronal recruitment, and in vitro measurements have confirmed that indeed a pulsating high frequency stimulation pulse will lead to neuronal recruitment.

In Fig. 7.3b it is illustrated that the proposed system is able to operate in multichannel mode without the need to duplicate the dynamic power supply or the inductor. The operating principle is shown in Fig. 7.4b for two channels. The high frequency pulses are delivered in an alternating fashion to both channels. Despite the fact that it will take twice as much time to deliver the same amount of charge to both loads, both channels will be activated simultaneously. Also note in Fig. 7.4b that it is possible to stimulate the channels independently with different amplitudes by adjusting the duty cycle for each channel individually, provided the dynamic supply operates in discontinuous mode.

Note that neuronal recruitment depends on the amount of charge injected (as described by the strength–duration curve [17]). This means that despite the fact that the average stimulation current is halved when two channels are operated simultaneously, the amount of charge delivered to the load remains the same as compared to single channel operation.

The power efficiency of the proposed dynamic power supply stimulator ideally does neither depend on α nor on the number of independent channels that are activated simultaneously. Instead it depends on the power efficiency of the dynamic stimulator. In the next section the system design of a high frequency dynamic stimulator is discussed.

7.2 System Design

7.2.1 High Frequency Dynamic Stimulator Requirements

The input of the system is assumed to be a Li-ion battery that is typically used in implantable systems, for which the nominal voltage is around $V_{in} = 3.5\,\text{V}$. The system is designed to connect to electrode leads that are used for deep brain and peripheral nerve stimulation. Platinum electrodes manufactured by St. Jude Medical are used that are ring shaped, have an area of approximately $14\,\text{mm}^2$, and for which it was assumed that $100\,\Omega < R_{load} < 1\,\text{k}\Omega$. The choice for these electrodes is not fundamental: the system can be designed to operate with other types of electrodes as well.

Stimulation amplitudes for these types of electrodes in commercial stimulators are reported up to 10 mA [18, 19]. This means that $V_{out} = I_{stim}R_{load}$ requires both up- and down-conversion with respect to V_{in} over the full range of R_{load}, which means that a buck-boost topology is needed for the dynamic stimulator. To avoid negative output voltages which complicate substrate biasing, it was chosen to use a forward buck-boost topology, which is shown as part of the total system in Fig. 7.5.

The switching frequency of the dynamic power supply determines the resolution for the pulse width. When N channels are active, the pulse width of each channel can be controlled in step sizes of N/f_{sw} seconds with f_{sw} being the switching frequency. It was chosen to have $f_{sw} = 1\,\text{MHz}$ such that the maximum step size is $8\,\mu\text{s}$ when all 8 channels are active.

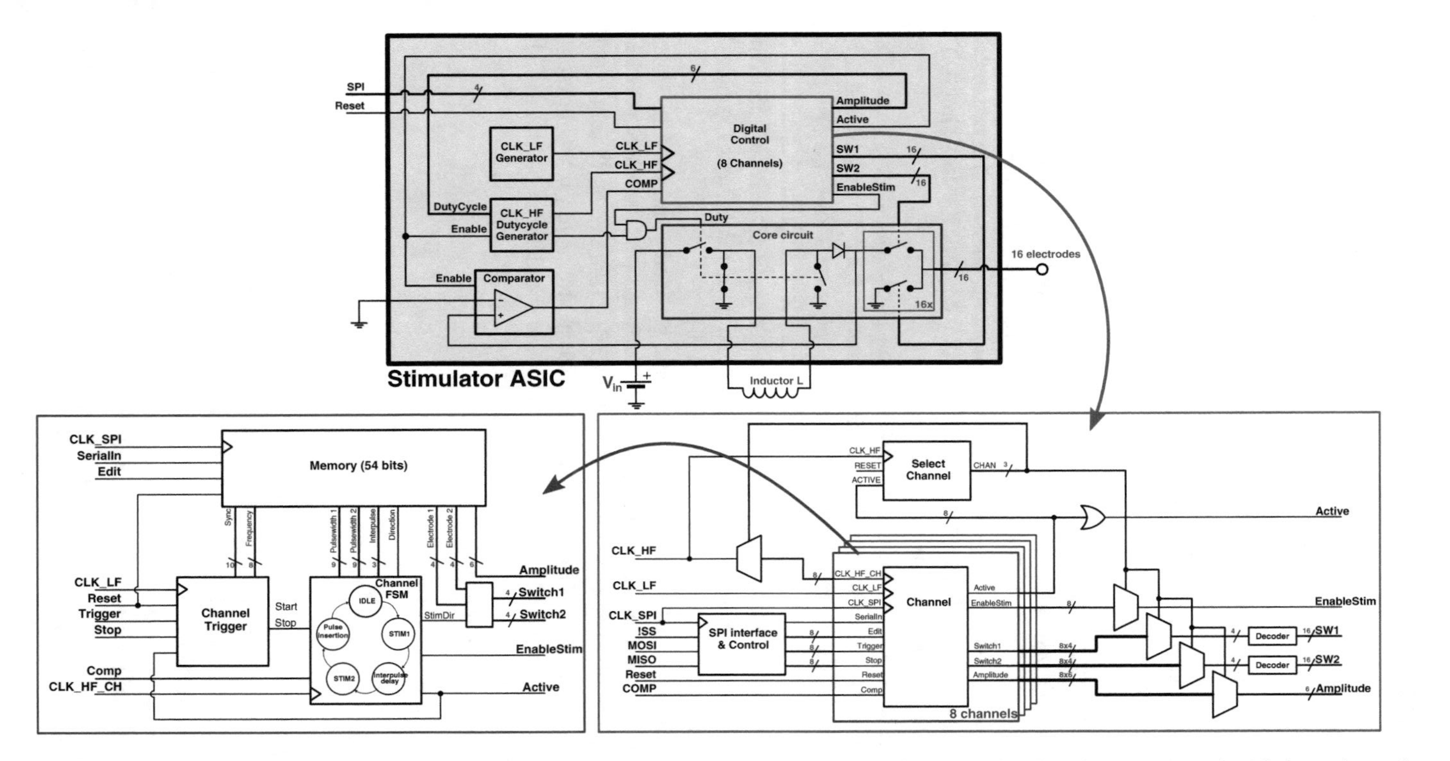

Fig. 7.5 System design overview. The core of the circuit is formed by a forward buck-boost dynamic supply that directly connects to the 16 electrodes at the output. A digital module, running on two clocks, controls the system and for which a detailed overview is given in the *green box*. Here the *Select_Channel* block selects out of the 8 channels the active ones and routes the input and output signals accordingly. The *red box* shows the functionality of a single stimulation channel

7.2.2 General System Architecture

The system design of the stimulator is depicted in Fig. 7.5. The forward buck-boost converter connects to 2×16 switches that make it possible for the user to select for each stimulation channel one electrode to be the anode and one to be the cathode. The digital control block generates all the necessary control signals to make the stimulator work. It implements 8 channels that have independent stimulation parameters, such as amplitude, pulsewidth, frequency, and the electrodes that are used. Each channel can be configured and controlled individually via an SPI interface.

The control block uses two clocks. The low frequency clock signal *CLK_LF* with $f_{clk\_lf} = 1$ kHz is always active and is used to trigger stimulation patterns. The high frequency clock *CLK_HF* with $f_{clk\_hf} = f_{sw} = 1$ MHz is used to control the core circuit and is only active when one or more stimulation channels are active. The signal *DUTY* is a duty cycled version of *CLK_HF* that controls the core circuit. The duty cycle is set by the 6 bit *AMPLITUDE* signal.

The comparator is used for charge cancellation purposes. After a stimulation cycle is finished, the remaining voltage over the electrodes is measured and using pulse insertion [20] the charge is removed from the interface. Pulse insertion is easy to incorporate in the system, since the stimulation waveform is pulse shaped already.

A biphasic stimulation pulse is generated by reversing the signals $SW1$ and $SW2$ after the first stimulation phase such that the direction of the current through the electrodes is also reversed (H-bridge configuration [9]).

Note that multiple channels can share the same electrodes, making the system suitable for many electrode configurations such as mono-, bi-, or tripolar as well as more complicated schemes required for field steering techniques. To achieve this level of flexibility in a classical current source based stimulator, it would require $8 \times 16 \times 2 = 172$ switches between the 8 current sources and the 16 electrodes. This system architecture therefore reduces the number of required switches by a factor 8 (i.e., by the number of channels).

7.2.3 Digital Control Design

In the bottom left corner of Fig. 7.5 a simplified structure of the block responsible for controlling one stimulation channel is depicted. In the memory 54 bits are used to store the stimulation settings. The memory is loaded via a serial interface when the *EDIT* pin is enabled.

The *Channel_FSM* is a finite state machine (FSM) running on *CLK_HF*, where the basic functionality of a stimulation pulse is implemented, as depicted in Fig. 7.6. After a trigger pulse is received, the FSM loops through the stages of a biphasic stimulation scheme where the pulse durations and interpulse delays are obtained by counting the number of *CLK_HF* periods equal to the value from the memory.

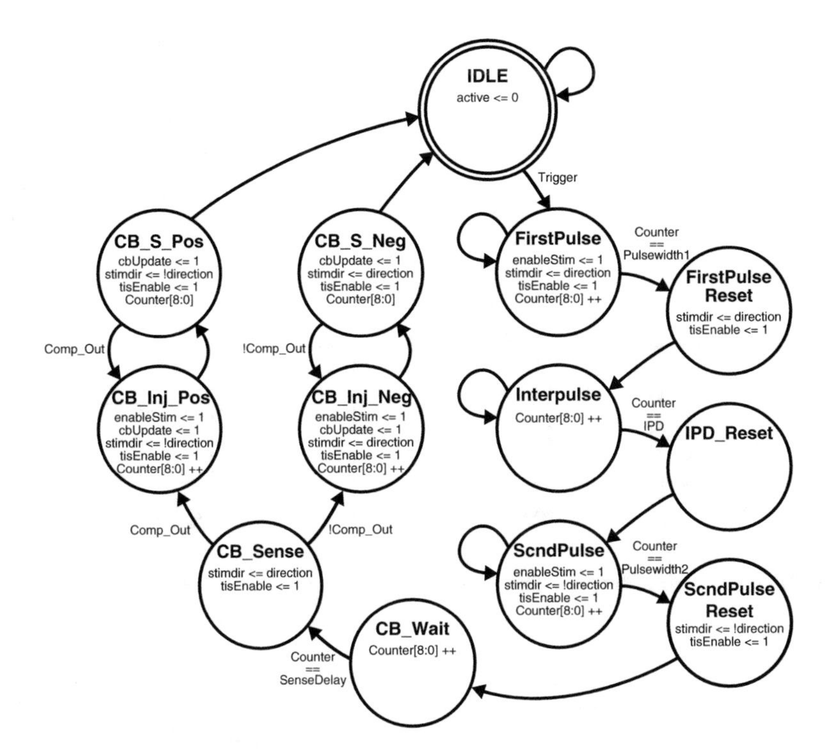

Fig. 7.6 FSM of the *Channel_FSM* block. When triggered it loops through the phases of a stimulation cycle consisting of the anodic pulse, the interpulse period, the cathodic pulse, and finally the charge balancing (CB) phase. The CB_wait state implements a delay that allows the electrode–tissue interface to settle to the equilibrium before the pulse insertion feedback mechanism (CB_sense) is started. Depending on the value of comp_out (the output of the comparator in Fig. 7.5) a pulse is inserted in the correct direction. The process is repeated until the comparator toggles (which means that the electrode voltage is zero again)

Afterwards the charge balancing is implemented using pulse insertion. The output signal *ACTIVE* is enabled when the FSM is not in the *IDLE* state, indicating that the channel is operating in a stimulation cycle. The *EnableStim* signal is used to enable the *DUTY* signal during the stimulation and charge cancellation phases.

After receiving a trigger or stop command, the *Channel_Trigger* block is able to start or stop a stimulation cycle in the *Channel_FSM* block in two different ways. When the frequency stored in the memory equals zero, the channel operates in 'single shot' mode: whenever it is triggered by an SPI command, a single stimulation sequence is generated. When the frequency is not zero, the channel operates in tonic mode: using *CLK_LF* a stimulation cycle is triggered periodically by counting the number of periods as specified by the value of frequency. Furthermore it is possible to align multiple channels by setting the *SYNC* value in the memory: upon triggering stimulation is delayed by a number of *CLK_LF* periods equal to the value in *SYNC*. In this way it is possible to accurately trigger multiple channels sequentially with only a single command.

Table 7.1 Commands used over the SPI interface to program the system

Command	Code
Edit channel	001
Trigger single channel	010
Stop single channel	011
Global trigger	100
Global stop	101

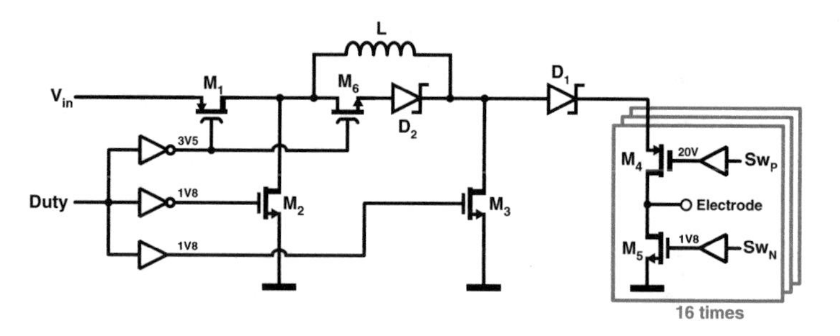

Fig. 7.7 Circuit implementation of the forward buck-boost dynamic supply

In the green block in Fig. 7.5 a simplified overview is given of the complete digital control system of the stimulator. At the core are the 8 channels. The *Select_channel* block is an FSM that keeps track of which channels are currently active. Using (de)multiplexers the inputs and outputs are routed to and from the active channel. If multiple channels are active at the same time, the *Select_channel* block alternates between the active channels.

The *SPI interface & control* block forms the interface with the outside world. The system can be configured with the commands shown in Table 7.1. The Edit Channel command is followed by a 3 bit word to select the channel and subsequently a 42 bit word is sent that contains the data to be stored in the memory of the channel (some of the Least Significant Bits in the 54 bits channel memory have default values). The Trigger and Stop Single Channel commands are also followed by a three bit code that selects the channel to be triggered or stopped. The global trigger and stop command are not followed by more bits since they affect all channel simultaneously.

7.3 Circuit Design

7.3.1 *Dynamic Stimulator*

The forward buck-boost topology from Fig. 7.5 was implemented as shown in Fig. 7.7. Transistors M_1, M_2, and M_3 form the dynamic supply switches, and M_4 and M_5 are the switches connecting the electrodes. Schottky diode D_1 is used to avoid oscillations in the load, while diode D_2 and switch M_6 are used to avoid oscillations in the inductor. All the gates of the transistors are driven with drivers that include level converters with appropriate voltages.

7.3.1.1 Choice of L

By using a first order Taylor approximation for the charging current of the inductor during the δT interval (δ being the duty cycle and $T = 1/f_{sw}$), the peak current in the inductor is $I_{peak} = V_{in}\delta T/L$ and the energy in the inductor will be $E_L = 0.5V_{in}^2\delta^2T^2/L$. In the ideal case all this energy is transferred to the load, leading to an average current $I_{avg} = \sqrt{E_L/(R_{load}T)}$. By combining these equations, an expression for the ratio $H = I_{avg}/I_{peak}$ is found:

$$H = \frac{I_{avg}}{I_{peak}} = \sqrt{\frac{L}{2TR}} \rightarrow L = 2RT\frac{I_{avg}^2}{I_{peak}^2} \tag{7.3}$$

This ratio should not become too small in order to prevent high values of I_{peak}. Therefore, for a given minimum H, the minimum value of L is determined.

The maximum value of the inductor is determined by the fact that the system is required to operate in discontinuous mode. For a given maximum duty cycle $\delta_{max} = 0.5$, the inductor needs to be discharged within the time frame $1 - \delta_{max}$. During discharging the system can be considered as a parallel RLC circuit consisting of the inductor L, the load R_{load}, and a parasitic capacitance C that is connected between the load and gnd. A conservative value of $C = 5$pF was chosen as an upper limit for the capacitance due to the bondpads, ESD protection, package pins, and other parasitic effects.

Conventional dynamic supplies use an output capacitor $C_{out} \gg C$, which will usually cause the parallel RLC circuit to be underdamped. Without C_{out} this 2nd order circuit is likely to be overdamped (which holds for $\zeta = \sqrt{T/(2RC)}H > 1$). The response of the system is: $V_{out} = A\exp(s_1t) + B\exp(s_2t)$. The real valued time constants $-s_1^{-1}$ and $-s_2^{-1}$ determine how fast the response decays. Taking the largest time constant $\tau = \max(-s_1^{-1}, -s_2^{-1})$, it was chosen to have $(1 - \delta_{max}) > 2\tau$. Calculating τ in terms of L gives:

$$L < -\frac{RT^2(\delta_{max} - 1)^2}{2(T\delta_{max} - T + 2RC)} \tag{7.4}$$

Equations (7.3) and (7.4) are plotted in Fig. 7.8 for maximum and minimum load conditions. Based on this figure it was chosen to have $L = 22\,\mu$H for which $0.105 < H < 0.33$ over the full range of R_{load}.

7.3.1.2 Conduction and Switching Losses

The sizing of transistors M_1–M_5 is important to find a trade-off between the conduction and switching losses. To analyze the response of the system including both conduction and switching losses, the circuit as depicted in Fig. 7.9 is analyzed, which includes the most dominant parasitic elements. During the charging phase S_2

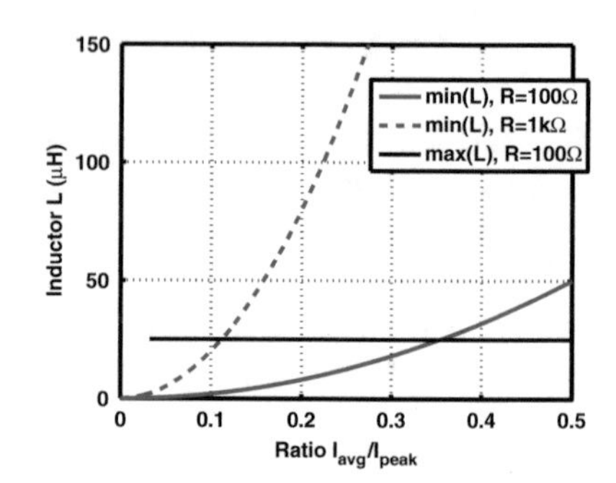

Fig. 7.8 Maximum and minimum values of L for different load conditions as a function of $H = I_{avg}/I_{peak}$

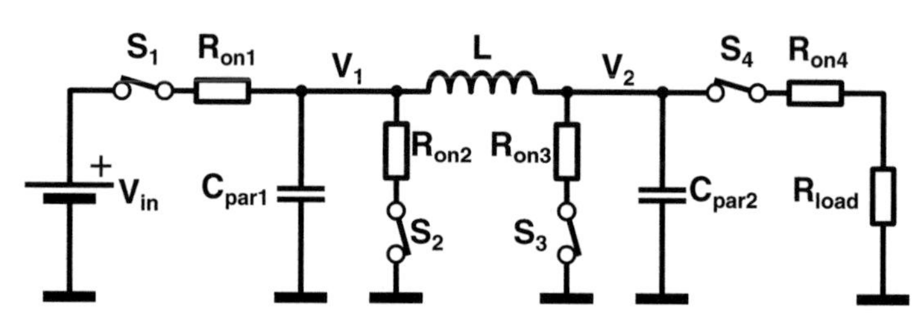

Fig. 7.9 Forward buck-boost converter circuit including conduction (R_{on}) and switching (C_{par}) losses

and S_4 are opened and can be removed from the circuit. Using Kirchhoff's current law (KCL) the following equations are obtained:

$$\frac{v_1 - V_{dd}}{R_{on1}} + C_{par1}\frac{dv_1}{dt} + \frac{1}{L}\int (v_1 - v_2)dt = 0 \tag{7.5a}$$

$$\frac{v_2}{R_{on2}} + C_{par2}\frac{dv_2}{dt} + \frac{1}{L}\int (v_2 - v_1)dt = 0 \tag{7.5b}$$

By substituting Eq. (7.5b) into (7.5a) the following third order differential equation can be found:

$$LC_{par1}C_{par2}\frac{d^3v_2}{dt^3} + \left[\frac{LC_{par2}}{R_{on1}} + \frac{LC_{par1}}{R_{on2}}\right]\frac{d^2v_2}{dt^2} + \left[\frac{L}{R_{on1}R_{on2}} + C_{par1} + C_{par2}\right]\frac{dv_2}{dt} + \left[\frac{1}{R_{on1}} + \frac{1}{R_{on2}}\right]v_2 = \frac{V_{dd}}{R_{on1}} \tag{7.6}$$

The roots s_1, s_2, and s_3 of the characteristic cubic equation can be found and by subsequently solving the particular solution, the following form for $v_2(t)$ is obtained:

$$v_2(t) = \mathrm{Re}\left\{K_1\exp(s_1t) + K_2\exp(s_2t) + K_3\exp(s_3t) + \frac{V_{dd}R_{on2}}{R_{on1} + R_{on2}}\right\} \tag{7.7}$$

Here K_1, K_2, and K_3 are found by solving the equations for the initial conditions $v_2(0) = 0\,\mathrm{V}$, $\frac{dv_2(0)}{dt} = 0$, and $\frac{dv_2^2(0)}{dt^2} = 0$, meaning that it is assumed that at the beginning of a new stimulation cycle there is no energy left in the dynamic components:

$$K_1 = \frac{-Zs_2s_3}{(s_1-s_2)(s_1-s_3)} \tag{7.8a}$$

$$K_2 = \frac{Zs_1s_3}{(s_1-s_2)(s_2-s_3)} \tag{7.8b}$$

$$K_3 = \frac{-Zs_1s_2}{s_1s_2-s_1s_3-s_2s_3+s_3^2} \tag{7.8c}$$

Here $Z = \frac{V_{dd}R_{on2}}{R_{on1}+R_{on2}}$. A very similar approach can be used to solve the circuit during the discharge phase. Now S_1 and S_3 are opened and can be removed from the circuit. Again by using the KCL and by using substitution, the following differential equation can be found:

$$LC_{par1}C_{par2}\frac{d^3v_2}{dt^3} + \left[\frac{LC_{par2}}{R_{on3}} + \frac{LC_{par1}}{(R+R_{on4})}\right]\frac{d^2v_2}{dt^2} + \tag{7.9}$$

$$\left[\frac{L}{R_{on3}(R+R_{on4})} + C_{par1} + C_{par2}\right]\frac{dv_2}{dt} + \tag{7.10}$$

$$\left[\frac{1}{R_{on3}} + \frac{1}{R+R_{on4}}\right]v_2 = 0 \tag{7.11}$$

Solving again for the roots s_1, s_2, and s_3 of this equation leads to the following expression:

$$v_2(t) = \mathrm{Re}\left\{K_1\exp(s_1t) + K_2\exp(s_2t) + K_3\exp(s_3t)\right\} \tag{7.12a}$$

$$K_1 = \frac{Z - Ys_2 - Ys_3 + Xs_2s_3)}{(s_1-s_2)(s_1-s_3)} \tag{7.12b}$$

$$K_2 = \frac{-(Z - Ys_1 - Ys_3 + Xs_1s_3)}{(s_1-s_2)(s_2-s_3)} \tag{7.12c}$$

$$K_3 = \frac{Z - Ys_1 - Y_s2 + Xs_1s_2}{s_1s_2 - s_1s_3 - s_2s_3 + s_3^2} \tag{7.12d}$$

The following constants are defined: $X = v_2(0)$, $Y = \frac{dv_2(0)}{dt} = C_{par2}^{-1}\left(I_L(0) - \frac{V_2(0)}{Ron4+R}\right)$, and $Z = \frac{dv_2^2(0)}{dt^2} = (LC_{par2})^{-1}\left(V_1(0) - V_2(0) - \frac{LY}{Ron4+R}\right)$, all of them determined by the initial conditions $v_1(0)$, $I_L(0)$, and $v_2(0)$ that are set by the corresponding values at the end of the charging phase. Using the expressions for $v_2(t)$, the expressions for $I_L(t)$ and $v_1(t)$ follow from the circuit.

The energy delivered by the source during the charging phase is described by the following equation: $E_s = \int V_{in}(V_{in} - v_1(t))/R_{on1}dt$. The energy dissipated in the

Table 7.2 Transistor sizes for the circuit in Fig. 7.7

Transistor	Size (W/L)
M_1	4800 μm/600 nm
M_2	1440 μm/200 nm
M_3	1680 μm/200 nm
M_4	1000 μm/600 nm
M_5	640 μm/200 nm

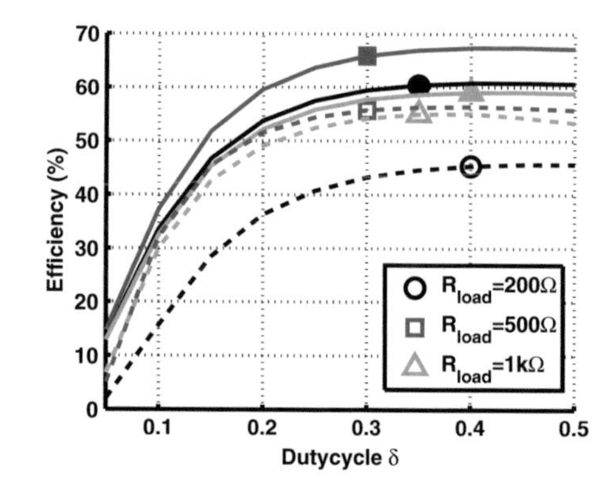

Fig. 7.10 Calculated (*solid lines*) and simulated (*dashed lines*) power efficiency of the dynamic stimulator circuit. Simulations include losses due to conduction, switching, gate drivers, bondpads, and non-ideal inductor

source during the discharge phase is found as $E_l = \int v_2^2(t)R/(R + R_{on4})^2dt$. The total power efficiency is now calculated as $\eta_{total} = E_l/E_s$.

The values of R_{on} and C_{par} can be found based on the parasitic components of the specific transistors as a function of their size. The bondpads, ESD circuitry, and package pins are again assumed to add an additional 5 pF to C_{par}. By iteratively evaluating the equations for various transistor sizes, a trade-off is found between conduction and switching losses. The calculated efficiency for various loads as a function of the duty cycle for the chosen transistor dimensions (see Table 7.2) is shown in Fig. 7.10.

Using the obtained transistor sizes, the circuit from Fig. 7.7 is simulated including gate driver and level converter circuitry. An inductor model that includes losses based on realistic inductors (Epcos 22 μH inductor with series resistance $R_s = 70\,\text{m}\Omega$ and parallel capacitance $C_p = 3.75\,\text{pF}$) is included. The simulation results are depicted in Fig. 7.10 as well. The efficiency is degraded with respect to the calculations due to three effects. First the implementation of the various components such as the drivers and the inductor introduces power losses. Second the voltage drop over diode D_1 was not accounted for in the previous calculations. Third, V_{gs} of transistor M_4 depends on the output voltage, because the gate is connected to *gnd* (it is not bootstrapped). For low output voltages (corresponding to low R_{load} and/or low δ), the second and third effects become dominant.

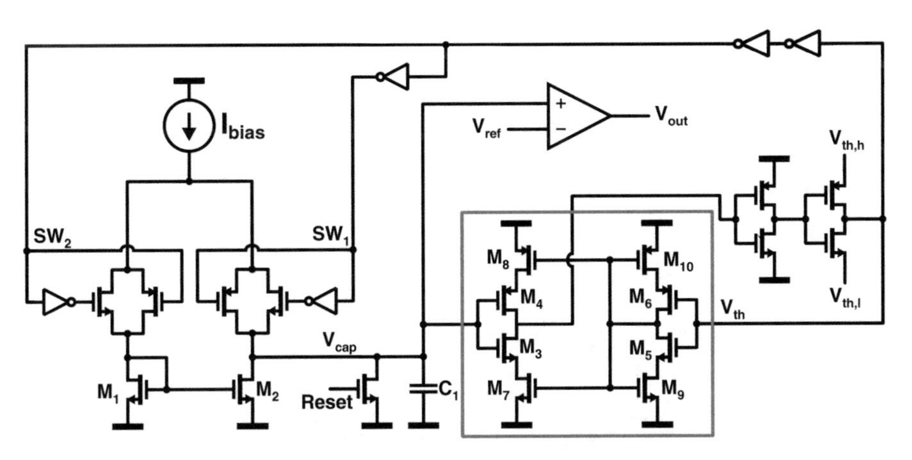

Fig. 7.11 Relaxation oscillator circuit used to generate the (duty cycled) clock signals. The circuit uses a Schmitt trigger that is implemented using a threshold compensated inverter, which is highlighted in the *green box*. The comparator is used to generate the *CLK_HF* and *DUTY* signals

7.3.2 Clock and Duty Cycle Generator

The clock signals *CLK_LF*, *CLK_HF*, and *DUTY* are all generated using the relaxation oscillator that is depicted in Fig. 7.11. The circuit uses a threshold compensated inverter [21] to implement a Schmitt trigger as was introduced in [22].

The bias current I_{bias} is used to charge capacitor C_1 by enabling the right-hand side transmission gate using SW_1. Once V_{cap} reaches the first threshold of the Schmitt trigger, SW_1 is opened and SW_2 closes, which causes the current direction through C_1 to reverse via current mirror M_1–M_2. This causes the voltage of C_1 to decrease again, until the second threshold of the Schmitt trigger is reached, which causes SW_1 and SW_2 to return to their original state, completing a clock period.

The threshold compensated inverter is highlighted in the box in Fig. 7.11. M_3 and M_4 form the inverter for which the threshold voltage is set by biasing M_7 and M_8 using the V_{th} signal. M_5 and M_6 are copies of M_3 and M_4 and generate the required biasing via the feedback loop comprising M_9 and M_{10}. The two thresholds of the Schmitt trigger are realized by switching V_{th} between $V_{th,l}$ and $V_{th,h}$ as shown in the figure.

The static current consumption through the branch M_{10}–M_6–M_5–M_9 is minimized by two mechanisms [22]. First of all V_{th} is chosen to be close to either V_{dd} or *gnd*, which switches off M_9 or M_{10}, respectively, and which leads to a large amplitude for V_{cap}. Second of all the length of M_9 and M_{10} can be increased, while the resulting V_{th} due to mismatch with M_7 and M_8 is very small.

The *CLK_LF* generator uses the SW_1 signal to obtain the output signal *CLK_LF*. The bias current for this block is 10 nA and the average simulated power consumption (including the $V_{th,l}$ and $V_{th,h}$ references) is 1.33 μW.

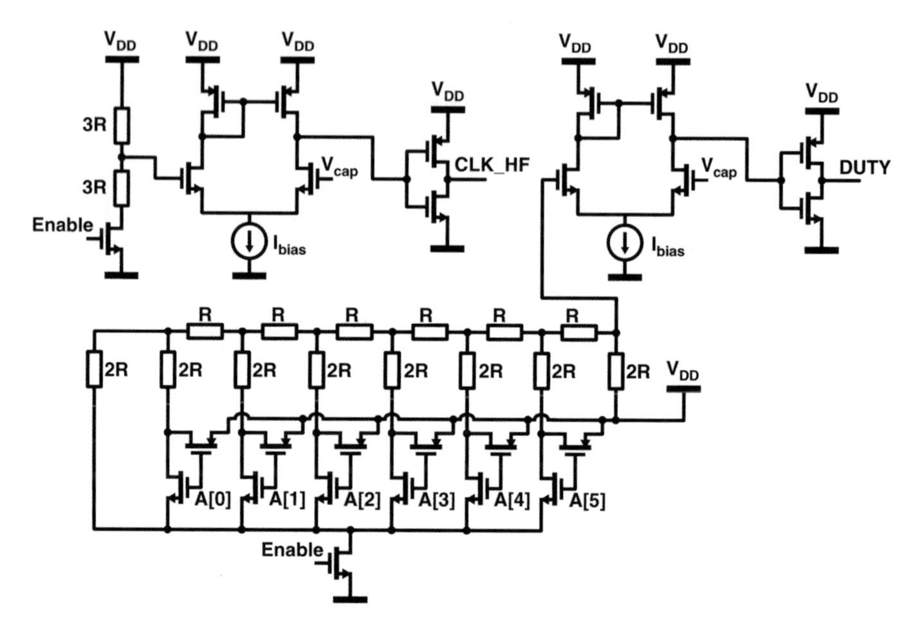

Fig. 7.12 DAC and comparator design used to generate the *CLK_HF* and *DUTY* outputs of the high frequency generator. The signal V_{cap} is connected to the same signal from Fig. 7.11

The 1 MHz duty cycled generator uses the triangular waveform of V_{cap} to generate a duty cycled signal by using the comparator as shown in Fig. 7.11. The value of V_{ref} is set using a digital to analog converter (DAC) using a standard R-2R structure ($R \approx 15\,\text{k}\Omega$) that is depicted in Fig. 7.12. As can be seen the *CLK_HF* signal is also derived from V_{cap} in order to make sure that *DUTY* is aligned with *CLK_HF*.

The average power consumption of the whole duty cycle generator over the full range of the DAC is simulated to be 176.2 μW. Notice that this block will only be active during stimulation and hence the average power consumption in a real situation will be much lower.

7.4 Experimental Results

The complete system has been implemented in 0.18 μm AMS H18 High Voltage technology. The digital control system is realized by synthesizing the Verilog description and occupies 0.25 mm^2. The total chip area of 3.36 mm^2 is pad limited and the layout with the various functional blocks highlighted is depicted in Fig. 7.13a. A microphotograph of the chip is given in Fig. 7.13b.

Besides V_{in} the system needs two more supply voltages. The supply voltage of the digital core as well as the clock generator blocks is $V_{dd,d} = 1.8$ V. The drivers of M_4 from Fig. 7.7 need $V_{dd,h} = 20$ V. The inductor used for the buck-boost system is the EPCOS B82464G4223M.

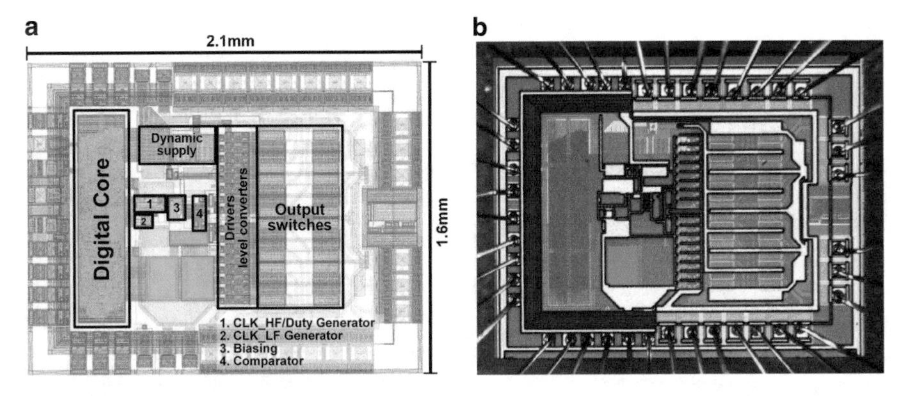

Fig. 7.13 In (**a**) a layout capture with the functional blocks highlighted is depicted. In (**b**) a microphotograph of the IC is shown

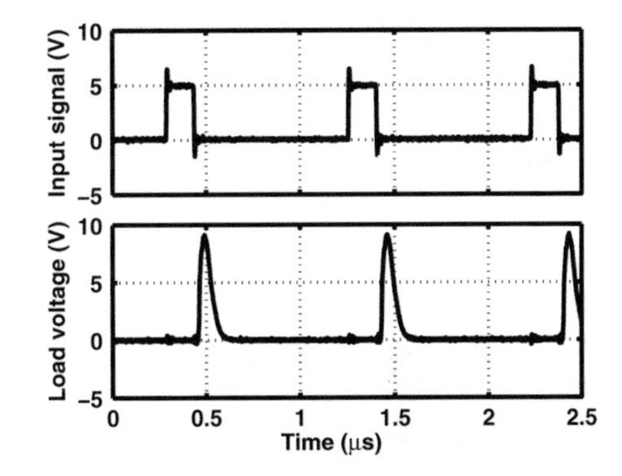

Fig. 7.14 Measurement results showing the transient operation on the system during a power efficiency measurement. The settings used for this measurement were $V_{in} = 3.5\,\text{V}$, $\delta = 0.15$, and $R_{load} = 1\,\text{k}\Omega$

7.4.1 Power Efficiency

The power efficiency of the buck-boost converter system is determined for various loads and stimulation intensities. The system is first configured to stimulate a resistive load continuously in one direction. In Fig. 7.14 an example is given of the waveform that is measured in this configuration.

Using a Keithley 6430 sourcemeter the average power supplied by the voltage sources is measured. The transient voltage V_{out} over the load is captured using a Tektronix TDS2014C oscilloscope and the average power is determined using Matlab by calculating $T^{-1} \int V_{out}^2 / R_{load} dt$.

The measured power efficiency of the dynamic converter (including the losses in the gate drivers) is depicted in Fig. 7.15a. As can be seen the measurements are in close correspondence with the simulation results. For high δ and high R_{load}, the efficiency goes down, because the output voltage is clipping to the supply voltage $V_{dd,h}$.

In Fig. 7.15b the power efficiency of the dynamic stimulator is measured for varying V_{in} in case of $R_{load} = 500\,\Omega$. As can be seen, the power efficiency of the system continues to be relatively high, although the available output power decreases for lower V_{in}.

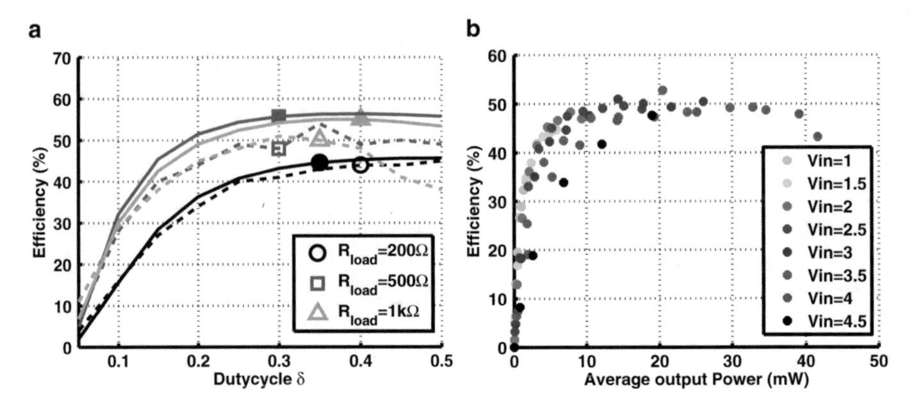

Fig. 7.15 Measurement results showing the power efficiency of the dynamic stimulator for various loads and duty cycles. In (**a**) the *dashed lines* are the measurement results, while the *solid lines* are the simulation results. In (**b**) the power efficiency for a 500 Ω load is depicted for various values of V_{in}

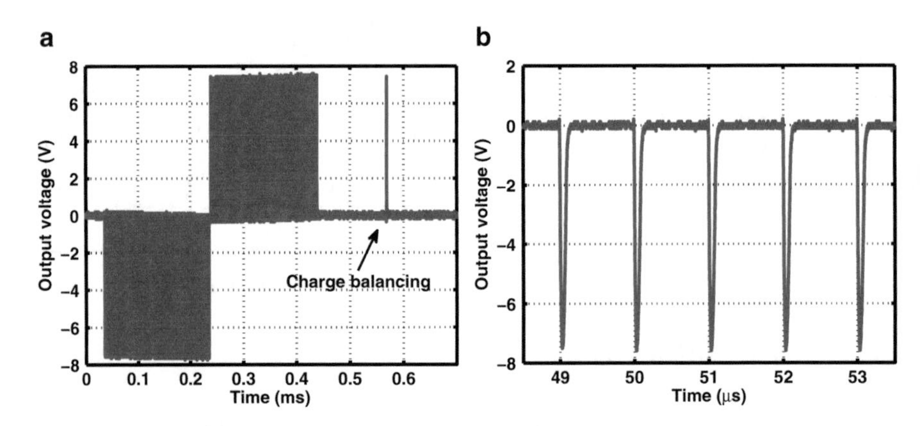

Fig. 7.16 Measured biphasic stimulation waveform using $t_{stim} = 200\,\mu\text{s}$ and $\delta = 0.17$ over a load of $R_{load} = 560\,\Omega$ and $C_{dl} = 1\,\mu\text{F}$. In (**b**) a detail of (**a**) is provided to show the shape of the individual pulses

7.4.2 Biphasic Stimulation Pulse

First a single channel is configured for a biphasic stimulation waveform with $\delta = 0.17$. The pulse width is 200 μs and the repetition rate is 3.92 Hz. The load is modeled using $R_{load} = 560\,\Omega$ and $C_{dl} = 1\,\mu\text{F}$.

The resulting waveform is depicted in Fig. 7.16a. The biphasic stimulation waveform has the expected shape and in Fig. 7.16b a detail of the stimulation waveform is given at the beginning of the first stimulation phase. After the biphasic stimulation pulse is finished, pulse insertion is used to remove the remaining charge from C_{dl}.

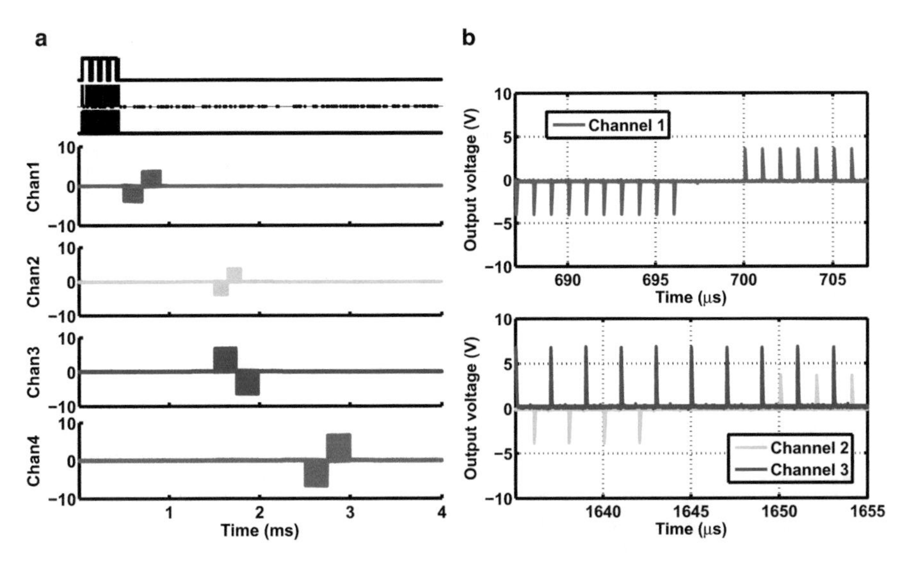

Fig. 7.17 Measurement results showing the multichannel operation of the system. In (**a**) the programming phase using SPI is shown for $t < 0.5$ ms, while afterwards 4 independent channels are shown. In (**b**) a detail is given for channel 1 and channel 2/3

7.4.3 Multichannel Operation

In Fig. 7.17a multichannel operation is demonstrated by activating four channels simultaneously. During the first 500 μs the SPI interface loads the stimulation settings for each channel individually and subsequently a single 'trigger global' command is given. Channel 1 starts immediately (sync = 0), channels 2 and 3 start 1 ms later (sync = 1), while channel 4 starts 2 ms later (sync = 2).

The detailed plots in Fig. 7.17b show that the simultaneous multichannel operation is working as expected: when two channels are active simultaneously, the pulses are alternately injected in each channel. Moreover it is seen that two channels can be stimulated simultaneously with opposite polarity, different pulsewidths, and amplitudes.

The power efficiency of the system in multichannel mode is compared to a constant current source with adaptive supply voltage. The first channel of the system is configured with a biphasic stimulation pulse with $\delta = 0.4$ through a load of $R_{load} = 500\,\Omega$. According to the measured average power in the load, this corresponds to an average $I_{stim} = 6.9$ mA. Additional channels are connected with $R_{load} = 200\,\Omega$. This degree of impedance variation can be caused by variances in the thickness of encapsulation tissue [13] and is not uncommon in clinical settings [23]. For each channel the same average I_{stim} is used, which corresponds to $\delta = 0.15$. The power efficiency of the proposed system in this configuration is shown in Fig. 7.18.

The equations from Sect. 7.1.1 with $\eta_{supply} = 80\,\%$ and $V_{compl} = 300$ mV are used to determine the power efficiency of a realistic adaptive supply constant current

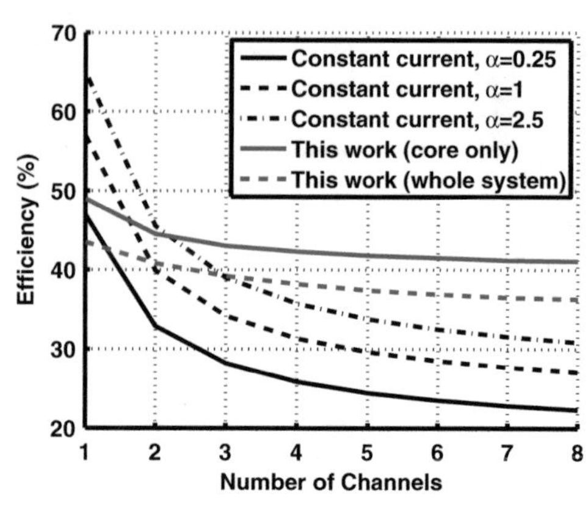

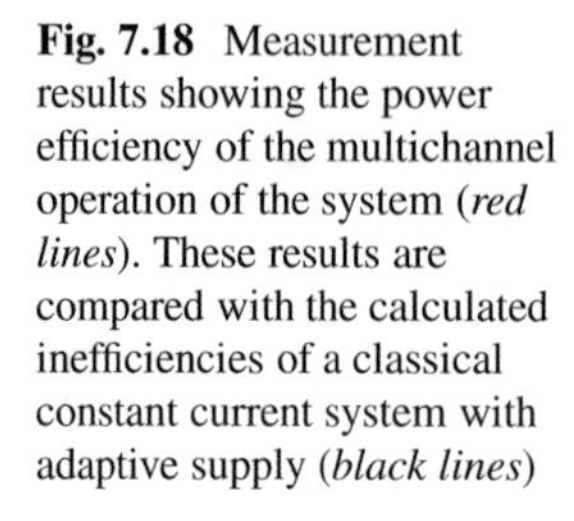
Fig. 7.18 Measurement results showing the power efficiency of the multichannel operation of the system (*red lines*). These results are compared with the calculated inefficiencies of a classical constant current system with adaptive supply (*black lines*)

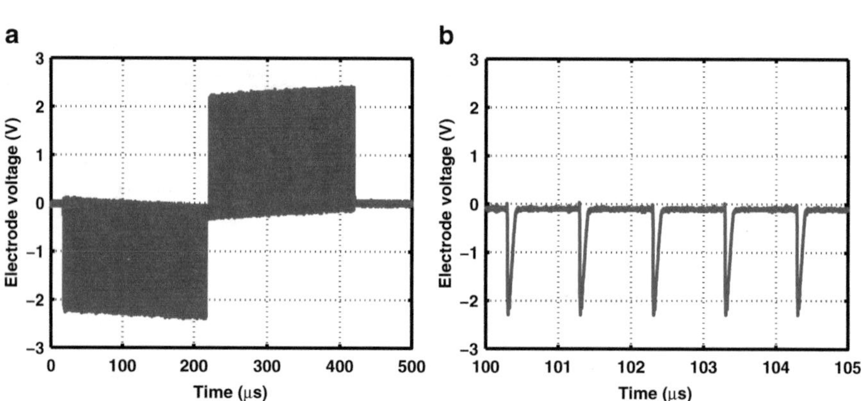

Fig. 7.19 Measured stimulation waveform for DBS electrodes submerged in a PBS solution bath. In (**b**) a detail is provided of (**a**) to see the shape of the individual pulses

stimulator in this situation for various values of α. As can be seen from Fig. 7.18, the system proposed in this work outperforms the adaptive power supply stimulator when operated using 2 or more channels. For low values of α this improvement can be as large as 200 %.

7.4.4 PBS Solution Measurements

The response of the system connected to electrodes in a Phosphate Buffered Saline (PBS) solution is measured. The platinum ring shaped electrodes with area 14 mm^2 as discussed earlier are submerged in a PBS solution bath. For the stimulation settings it was chosen to have $\delta = 0.15$ and $t_{stim} = 200\,\mu s$. The resulting electrode voltage is shown in Fig. 7.19 and looks very similar to the result from the series RC model.

7.4.5 Discussion

The power efficiency of the system can be further improved, especially by reducing the losses in D_1 and M_4 from the circuit in Fig. 7.7. These devices could be combined in a single transistor that is operated with bootstrapping, which will significantly reduce the losses for low output voltages. Moreover, the power efficiency of the duty cycle generator can be improved by designing a more power efficient DAC topology.

The system currently still needs two external supplies: one low supply for the digital control and one high supply for the HV switches. Future implementations can integrate the required voltage converters. The power needed from $V_{dd,h}$ heavily depends on the stimulation settings, but is generally relatively low, which makes it possible to integrate a charge pump for this purpose.

The number of electrodes connected to the system in its current form cannot be increased without penalties. Each additional electrode requires an additional switch M_4 in Fig. 7.7, which increases the parasitic capacitance at this node, increasing the switching losses in the circuit. One possible way to overcome this is to design more sophisticated switch array configurations that aim to minimize the capacitive load.

Another limitation of the designed prototype is that it currently is operated in open loop: there is no control over the amount of injected charge, other than by controlling the duty cycle. Future implementations can benefit from including a feedback mechanism that controls the charge delivered to the load and can compensate for variations in, for example, V_{in} and L.

One advantage of the proposed system that has not been addressed yet is that there is no driver transistor that connects V_{dd} directly to the electrodes. Such a driver transistor is found in current source based implementations and introduces a single fault failure mode for the device: when this device is shorted, a large current will flow through the electrodes. In the proposed system V_{in} connects to the electrode via multiple switches and hence this does not introduce a potential single fault failure mode.

It is likely that the proposed stimulation strategy can also be used for other kind of excitable tissue, such as muscle tissue. However, there is more research needed towards the effect of the high frequency current pulses on the tissue. In [24] the efficacy of this type of stimulation was shown, but little is known about the losses in the tissue and (long term) safety aspects. On the other hand it is interesting to see that the proposed stimulation principle mimics the natural working principle of neurons: the synaptic receptors continuously receive pulsating inputs that are integrated on the membrane surface of the dendritic tree. The pulsed stimulation has a similar peaking waveform.

7.5 Conclusions

This chapter has presented the realization of a neural stimulator system that uses an unfiltered dynamic supply to directly stimulate the target tissue. It is possible to operate the system with multiple independent channels that connect to an arbitrary electrode configuration, making the system well suited for current steering techniques. Furthermore, comprehensive control was implemented using a dual clock configuration that allows both autonomic tonic stimulation as well as single shot stimulation. Each channel can be configured individually with tailored stimulation parameters and multiple channels can operate in a synchronized fashion.

The system is shown to be power efficient, especially when compared with state-of-the-art constant current stimulators with an adaptive power supply that operate in multichannel mode. Efficiency improvements up to 200 % have been demonstrated.

References

1. Bonham, B.H., Litvak, L.M.: Current focusing and steering: modeling, physiology and psychophysics. Hear. Res. **242**(1–2), 141–153 (2008)
2. Noorsal, E., Sooksood, K., Xu, H., Hornig, R., Becker, J., Ortmanns, M.: A neural stimulator frontend with high-voltage compliance and programmable pulse shape for epiretinal implants. IEEE J. Solid State Circuits **47**(1), 244–256 (2012)
3. Lo, Y.K., Chen, K., Liu, W.: A fully-integrated high-compliance voltage SoC for epi-retinal and neural prostheses. IEEE Trans. Biomed. Circuits Syst. **7**(6), 761–772 (2013)
4. Chen, K., Yang, Z., Hoang, L., Weiland, J., Humayun, M., Liu, W.: An integrated 256-channel epiretinal prosthesis. IEEE J. Solid-State Circuit **45**(9), 1946–1956 (2010)
5. Veraart, C., Grill, W.M., Mortimer, T.: Selective control of muscle activation with a multipolar nerve cuff electrode. IEEE Trans. Biomed. Eng. **40**(7), 640–653 (1993)
6. Martens, H.C.F, Toader, E., Decré, M.M.J., Anderson, D.J., Vetter, R., Kipke, D.R., Bakker, K.B., Johnson, M.D., Vitek, J.K.: Spatial steering of deep brain stimulation volumes using a novel lead design. Clin. Neurophysiol. **122-3**, 558–566 (2011)
7. Valente, V., Demosthenous, A., Bayford, R.: A tripolar current-steering stimulator ASIC for field shaping in deep brain stimulation. IEEE Trans. Biomed. Circuits Syst. **6**(3), 197–207 (2012)
8. Sooksood, K., Noorsal, E., Bihr, U., Ortmanns, M.: Recent advances in power efficient output stage for high density implantable stimulators. 2012 IEEE Annual International Conference of the Engineering in Medicine and Biology Society (EMBS), pp. 855–858 (2012)
9. Williams, I., Constandinou, T.G.: An energy-efficient, dynamic voltage scaling neural stimulator for a proprioceptive prosthesis. IEEE Trans. Biomed. Circuits Syst. **7**(2), 129–139 (2013)
10. van Dongen, M.N., Serdijn, W.A.: A power-efficient multichannel neural stimulator using high-frequency pulsed excitation from an unfiltered dynamic supply. IEEE Trans. Biomed. Circuits Syst. (2014). http://ieeexplore.ieee.org/xpl/articleDetails.jsp?arnumber=6965660
11. Lee, H.N., Park, H., Ghovanloo, M.: A power-efficient wireless system with adaptive supply control for deep brain stimulation. IEEE J. Solid-State Circuits **48**(9), 2203–2216 (2012)
12. Randles, J.E.B.: Kinetics of rapid electrode reactions. Discuss. Faraday Soc. **1**, 11–19 (1947)
13. Butson, C.R., Maks, C.B., McIntyre, C.C.: Sources and effects of electrode impedance during deep brain stimulation. Clin. Neurophysiol. **117**(2), 447–454 (2006)

14. Cheung, T., Nuo, M., Hoffman, M., Katz, M., Kilbane, C., Alterman, R., Tagliati, M.: Longitudinal impedance variability in patients with chronically implanted DBS devices. Brain Stimul. **6**, 746–751 (2013)
15. Arfin, S.K., Sarpeshkar, R.: An energy-efficient, adiabatic electrode stimulator with inductive energy recycling and feedback current regulation. IEEE Trans. Biomed. Circuits Syst. **6**(1), 1–14 (2012)
16. van Dongen, M.N., Serdijn, W.A.: A switched-mode multichannel neural stimulator with a minimum number of external components. IEEE International Symposium on Circuits and Systems (ISCAS) (2013)
17. Malmivuo, J., Plonsey, R.: Bioelectromagnetism – Principles and Applications of Bioelectric and Biomagnetic Fields. Oxford University Press, New York (1995)
18. Kuncel, A.M., Grill, W.M.: Selection of stimulus parameters for deep brain stimulation. Clin. Neurophysiol. **115**(11), 2431–2441 (2004)
19. Slavin, K.V.: Peripheral nerve stimulation for neuropathic pain. Neurotherapeutics **5**(1), 100–106 (2008)
20. Sooksood, K, Stieglitz, T., Ortmanns, M.: An active approach for charge balancing in functional electrical stimulation. IEEE Trans. Biomed. Circuits Syst. **4**(3), 162–170 (2010)
21. Tan, M.T., Chang, J.S., Tong, Y.C.: A process-independent threshold voltage inverter-comparator for pulse width modulation applications. Proceedings of IEEE International Conference on Electronics, Circuits and Systems, vol. 3, pp. 1201–1204 (1999)
22. van Dongen, M.N., Serdijn, W.A.: Design of a low power 100 dB dynamic range integrator for an implantable neural stimulator. IEEE Biomedical Circuits and Systems Conference (BioCAS), pp. 158–161 (2010)
23. Sillay, K.A., Chen, J.C., Montgomery, E.B.: Long-term measurement of therapeutic electrode impedance in deep brain stimulation. Neuromodulation **13**(3), 195–200 (2010)
24. van Dongen, M.N., Hoebeek, F.E., Koekoek, S.K.E., De Zeeuw, C.I., Serdijn, W.A.: High frequency switched-mode stimulation can evoke postsynaptic responses in cerebellar principal neurons. Front. Neuroengineering **8**(2) (2015)

Chapter 8
Conclusions

This book has taken a multidisciplinary approach for the design of neural stimulators. Electrophysiological and electrochemical principles that govern the working principles of functional electrical stimulation were combined with electrical engineering aspects to introduce stimulation concepts that differ from the traditional constant current/voltage approach and that offer advantages in terms of efficiency and/or safety. In Chap. 2 it was shown that electrical stimulation can be considered at three different levels: the electrode level, the tissue level, and the neuronal level.

In Chap. 3 the electrode level is considered in relation with the safety of neural stimulation. The voltage over the electrode–tissue interface should be limited in order to prevent harmful electrochemical reactions. There are various methods available to prevent charge accumulation at the interface, such as biphasic charge balanced stimulation, coupling capacitors, and electrode shortening. This chapter first explored the consequences of the use of coupling capacitors. It was found that, in contrast to what most existing studies suggest, coupling capacitors do not improve the charge balancing at the electrode interface. Furthermore they introduce an offset voltage that, depending on the stimulation settings, might reach potentially dangerous levels. Therefore, special attention should be paid to eliminate the risks of such an offset when the use of coupling capacitors is required for other safety reasons. Furthermore, this chapter has explored the use of a feedforward charge balancing scheme that aims to bring the interface back to equilibrium after a stimulation cycle by exclusively balancing the capacitive (reversible) currents.

In Chap. 4 the tissue and neuronal level of neural stimulation is considered to explore the concept of high frequency duty cycled stimulation. Using modeling that included the dynamic properties of both the tissue material as well as the axon membrane it was found that high frequency stimulation signals can recruit neurons in a similar fashion as classical constant current stimulation. The response of Purkinje cells due to stimulation in the molecular layer was measured for both classical and switched-mode stimulation. The measurements confirmed the modeling in showing that switched-mode stimulation can induce neuronal activation

M. van Dongen, W. Serdijn, *Design of Efficient and Safe Neural Stimulators*,
Analog Circuits and Signal Processing, DOI 10.1007/978-3-319-28131-5_8

and that both the duty cycle and the stimulation voltage are effective ways to control the intensity of the stimulation. Care has to be taken to avoid losses in the stimulation system that arise due to the use of a high frequency stimulation signal.

In the second part of the book the gears are switched to the electrical implementation of neural stimulators. Two stimulator designs for two different applications are presented.

The first stimulator design has focused on arbitrary stimulation waveforms, while charge cancellation is guaranteed. An accurate scaled copy of the stimulation current was generated in order to determine the charge contents of the stimulation signal. This allows for arbitrary waveforms and asymmetrical biphasic stimulation while charge cancellation is maintained. This is a useful feature for experimental setups in which the user wants full freedom over the stimulation pattern.

Both simulations and the discrete realization achieve a charge mismatch of several percent, depending on the waveform settings. When combined with electrode shortening, the circuit can be applied in practice when the stimulation rate is low enough to allow sufficient discharge during the shortening phase. A complete stimulation system was engineered to be used in animal experiments for tinnitus treatment, in which the electrical stimulation signal was combined with an auditory stimulation.

The second stimulator design implements the high frequency duty cycled stimulation as introduced in Chap. 4. The system is realized using an unfiltered dynamic supply that directly stimulates the target tissue. The focus in this design is on power efficiency, a low number of external components, and independent multiple channel operation. All these requirements are important for implantable neural stimulators that use current steering techniques, such as SCS, VNS, or DBS applications.

The mixed-signal IC implementation features a complete stimulator system, including comprehensive control that was implemented using a dual clock configuration that allows both autonomic tonic stimulation, as well as single shot stimulation. Each channel can be configured individually with tailored stimulation parameters and multiple channels can operate in a synchronized fashion. Each stimulation channel can connect to an arbitrary electrode configuration, making the system well suited for current steering techniques.

The system is shown to be power efficient when compared with state-of-the-art constant current stimulators with an adaptive power supply, especially when operated in multichannel mode. Efficiency improvements up to 200 % have been demonstrated. Furthermore, the system uses an inductor as its only external component, improving the level of integration with respect to existing stimulator systems that often also need one or more external capacitors.

Index

M. van Dongen, W. Serdijn, *Design of Efficient and Safe Neural Stimulators*,
Analog Circuits and Signal Processing, DOI 10.1007/978-3-319-28131-5

O

P

R

S

T

U

V

Printed in the United States
By Bookmasters